About the Author

PAUL KENGOR is the author of the *New York Times* extended-list bestseller *God and Ronald Reagan* as well as *God and George W. Bush* and *The Crusader.* He is a professor of political science and director of the Center for Vision and Values at Grove City College. He lives with his wife and children in Grove City, Pennsylvania.

GOD

and

HILLARY CLINTON

GOD

and

HILLARY CLINTON

a spiritual life

PAUL KENGOR, PH.D.

HARPER PERENNIAL

NEW YORK • LONDON • TORONTO • SYDNEY • NEW DELHI • AUCKLAND

HARPER ● PERENNIAL

FIRST HARPER PERENNIAL EDITION PUBLISHED 2008.

The Library of Congress has catalogued the hardcover edition as follows:
    Kengor, Paul.
    God and Hillary Clinton : a spiritual life / Paul Kengor.—1st ed.
       p.   cm.
    Includes bibliographical references and index.
    ISBN 978-0-06-113692-4 (hardcover)
       1. Clinton, Hillary Rodham—Religion. 2. Clinton, Hillary Rodham—Political and social views. 3. Presidents' spouses—United States—Biography. 4. Legislators—United States—Biography. 5. Governors' spouses—Arkansas—Biography. 6. United States—Politics and government—1993–2001. 7. Arkansas—Politics and government—1951– 8. Religion and politics—United States—Case studies. I. Title.
    E887.C55K455   2007
    328.73092—dc22
    [B]
    2007019430

    ISBN 978-0-06-118925-8 (pbk.)

08 09 10 11 12 ID/RRD   10 9 8 7 6 5 4 3 2 1

My prayer for you is that you come to understand
and have the courage to answer.
—*Mother Teresa*

Contents

Preface

Writing a book on the faith of Hillary Rodham Clinton has been a fascinating but frustrating process. On the latter, many of the sources contacted or consulted in the course of researching and writing this book had some type of problem. With my previous two faith-based subjects, Ronald Reagan and George W. Bush, I found that while pastors commonly respected a level of confidentiality concerning their congregation and were often unwilling to share personal material on the subjects, friends and associates were typically happy to talk. For Reagan in particular, many were eager to finally get the word out and set the record straight on the man's deep, unappreciated faith. As a man who is quite open about his faith, George W. Bush was also an easier subject.

Since I began writing this spiritual biography of Hillary Clinton, however, I encountered a real reluctance among some who knew her through church to talk about their experiences with her, fearing the information would be used against Hillary—even in instances when it was positive. Secondary sources were likewise problematic, as many biographers lionize or demonize Hillary and her husband, sometimes reporting sensational stories that—even when seemingly plausible—are based on unnamed or flawed sources. Much of this problematic research often does not use endnotes and frequently repeats the previous writers' unsourced material.

Moreover, many sources, as the Clintons themselves have complained, report explosive sexual innuendo—material that in certain cases has been confirmed, at times through the controversial work of official government investigators. This means that a biographer of the Clintons is sometimes forced to delve into the lurid in explaining them—uncharted waters to a college professor like me. Unfortunately, one cannot talk about the Clintons and their relationship, and particularly their spiritual relationship, without discussing this elephant in the bedroom, since it has been a dominant element of their marriage, their faith, and their very public lives.

And then there are the most problematic sources of all: the Clintons. Like them or not, the fact is that Hillary and Bill Clinton may be the most thoroughly political first couple in the history of American politics. More, they are leftists, once radical leftists, who have learned to campaign to the middle and to moderate stances on key swing issues in order to get elected. Of course, many politicians do this, but the Clintons do so more frequently than any politicians on the left or right, and as a result, they are not always fully forthcoming about what they believe, which makes reporting on them so sticky and, for the biographer, risky.

This was a hurdle I faced when teaching my students about President Bill Clinton. For most presidents I lectured on, whether Democrat or Republican, writing their beliefs on the chalkboard was a fairly straightforward process. With Bill Clinton, that was always a struggle because of that unique political element. *What does he truly believe?* It was a question similar to that famously expressed by an exasperated James Carville, Clinton's colorful "ragin' Cajun" political adviser, who banged the table and asked of his boss, "Where's the sacred ground?!" Or, as *Washington Post* reporter and talented Clinton biographer David Maraniss has put it, the "perpetual question" that hangs over Bill Clinton is the question of "what he really stands for."[1]

While Bill and Hillary are truly distinct politicians, what applies in this regard to Bill partly applies to Hillary as well. Indeed, Hillary often advised her husband to adopt this political approach and honed the tactic herself at the side of the master. Now she, too, holds elected office and seeks the highest office in the land, looking to position herself as a moderate, just as her husband did.

This challenge regarding the Clintons becomes even more difficult when trying to address the subject of religion. The 2004 election reaffirmed what many Democrats—Hillary foremost among them—now understand: A modern liberal campaigning for president must appear at least partly religious, or at least not unreligious or antireligious. The Democratic Party responded astutely, employing an impressive example of correctly reading the electorate. In 2006, the Democratic Party leadership captured the Congress in large part by running moderate, religious, at times even pro-life candidates, embodied by the likes of Bob Casey Jr. in Pennsylvania.

Hillary and her husband recognized the political potency of religion long before religiously minded voters helped George W. Bush win consecutive presidential elections. Both have unapologetically campaigned in churches more brazenly than politicians in either party, and at times (in Mrs. Clinton's case) in racially polarizing ways. The secular press—in an example of striking hypocrisy—has winked at the behavior, allowing the Clintons full immunity in pursuing this integration of church and state, in a manner that George W. Bush would not have been able to get away with, even if he so desired.

These Clintonian dashes to the middle for political positioning are a real problem for a biographer. Time and time again in researching and writing this book, I was left to conclude that on certain religious questions, the only sources who truly know the answer are God, Hillary, and possibly Bill Clinton. One source would say one thing while another source would say something else. Finally Hillary (or Bill) would weigh in with yet another version of an idea, creating a vague gray area of truth and leading to treacherous ground for a biographer.

That said, some things regarding Hillary Clinton and her faith are clear: Although no one can profess to know any individual's heart and soul, there seems no question that Hillary is a sincere, committed Christian and has been since childhood. The same applies to her husband, who admits that he is a sinner—as are all Christians, which is why they are Christians. Surely not even the most cynical right-winger would insist that Hillary and Bill were playing politics when they eagerly attended Sunday school as eight-year-olds. Hillary is a very liberal Christian, and would be categorized as part of the religious left, along with millions of Christian Americans—a designation that seems to have disappeared from the media's lexicon now that the secular press is obsessed with fears over the religious right.

The faith of both of the Clintons affects their lives and their politics. Mrs. Clinton has not been shy about incorporating her faith into the policies she advocates, and probably does so as much as well-known religious politicians on the right, such as the current president. If ever confronted with the charge from the usually vigilant church-state-separation media, she (and her husband) would be hard pressed to dodge the obvious assertion that they have at times exploited their religious faith for political purposes.

The greatest paradox of Mrs. Clinton's faith, the one that will hurt her most in her bid to attract those "values voters" she needs in 2008—including even a sliver of the churchgoing Catholics and Protestants who overwhelmingly went for Bush—is the way her Christian faith has reinforced her commitment to human rights when it comes to civil rights and children's rights and gay rights, but not the rights of unborn children, where her faith has given way to her ideology. There is no issue more impassioning to Senator Clinton than abortion rights, meaning that any serious analysis of her political-religious thinking is forced to devote a significant amount of attention to the subject, as she does herself in her own personal and professional life.[2]

Among those religious voters, the Reverend Jerry Falwell remarked at the September 2006 "Values Voters Summit": "I certainly hope that Hillary is the candidate [in 2008]. Because nothing will energize my [constituency] like Hillary Clinton. If Lucifer ran, he wouldn't."[3] Falwell's assessment is a measure of the uphill battle that Mrs. Clinton faces. She will not win Falwell supporters, but hopes to attract just enough religious moderates.

The faith of Hillary Clinton is a subject that requires close examination, featuring a lot of drama and twists and turns, conventional and unconventional, from her studies of the classic theologian John Wesley to strange moments of imaginary conversation with a deceased Eleanor Roosevelt. And though one's personal relationship with God is private—man or woman and his or her Maker—the subject in Hillary's case is of interest to all Americans because of its prominent place in her private and public life and because she has already begun to campaign vigorously for president as a religious Democrat—more so than any Democrat since Jimmy Carter.

What follows is a spiritual biography of this leading political figure, a chronology of the story of her complicated yet intricate Christian faith.

GOD

and

HILLARY CLINTON

CHAPTER I

Park Ridge Methodist

Hugh Ellsworth Rodham was tough as nails. Born in 1911, he grew up in the mining town of Scranton, Pennsylvania, and managed to get himself educated during the Great Depression by winning a football scholarship to Penn State University, where he studied physical education. Phys ed looked like a good choice for Hugh, and had he chosen that path he might have matched the image that many young men have of a high school gym teacher who barks out instructions and calls them "ladies," generally questioning their manhood until they successfully bean a classmate or two in the head with a dodgeball.

Hugh, however, did not follow that road. Instead, he graduated from Penn State with his bachelor's degree in education in 1935 and went to work in the mines—the expected course for the Rodhams of Scranton—before later joining his father in the notably less dismal Scranton Lace Company. Still not content with the gray mining and manufacturing town, he packed his bags and began hopping on and off freight cars all the way to Chicago, where he found employment selling curtains at the Columbia Lace Company. It was in that capacity that he spotted a young lady named Dorothy.

Eight years younger than Hugh, Dorothy Emma Howell had a disturbing childhood. Born to a fifteen-year-old mother and a seventeen-year-old father in Chicago in 1919, little Dorothy saw her parents divorce in 1927. Her mother, Della, sent eight-year-old Dorothy along with her three-year-old sister across the country by train on a four-day trek, reportedly with no adult accompaniment, to a small town near Los Angeles, where the children lived with a badgering, cruel grandmother who criticized the innocent girls' every move.

By the time Dorothy turned fourteen, she had found life in her grandmother's home intolerable. Without much ceremony, the young woman grabbed her one blouse, one skirt, and one sweater—her entire wardrobe at the time—and sought employment as a mother's helper for two children at a nearby home. The job paid $3 a week, but it also gave her room and board, an experience that gave Dorothy the chance to discover what love between parents and their children was supposed to look like. It was a literally life-changing experience for Dorothy, and years later Hillary would say that her "mother often told me that without that sojourn with a strong family, she would not have known how to care for her own home and children."[1]

As she worked to help the family, Dorothy continued to attend high school. The young girl loved to read and hoped somehow to attend college, but shortly after Dorothy's successful completion of high school, Della got in touch with her daughter. Della, who was still living in Chicago, had remarried, and according to Della, her new husband promised to pay for Dorothy to attend college back in Illinois. Eager to learn and aspiring to be a part of a family like the one she had worked for, Dorothy arrived "home" to find that Della, a weak basket case of a woman, had lied. The whole situation had been a cruel hoax to try and lure Dorothy back to Chicago so that she could work as a housekeeper for Della. Sadly, her mother could not have cared less about giving her an education.[2]

Despite her mother's attempt to put Dorothy to work, the young woman refused to be ensnared, opting instead to go off on her own once again. It did not take her long to find an apartment, and soon after she began searching for a low-paying office job to pay her rent. She was in the middle of her search, filling out an application for a position as a clerk-typist at a textile company, when she caught the eye of a traveling salesman named Hugh Rodham. That one glance was all it took, and the couple courted for a while before marrying in early 1942.

Hugh continued his sales job through the war years, but contributed his part to the war effort, serving his country as a trainer for navy recruits sent abroad to fight in the Pacific theater. In these efforts, he applied the same tenacity that had made him a successful competitor on the football field, barking orders at young men and forcing them to push their bodies to the brink. Hugh took great pride in this form of military service, and though he did not see combat or ever travel abroad to fight, he rose to the rank of chief petty officer in the navy.

When World War II ended, Hugh started a drapery-fabric business called Rodrik Fabrics in the Merchandise Mart in Chicago's Loop. By 1950, his company was thriving, and he was suddenly able to give Dorothy the comfort and stability she never had and much deserved. He paid cash for a two-story brick house situated on a corner lot between Elm and Wisner streets in the affluent Park Ridge suburb of Chicago. It was a defining move for the young couple, one that offered them the perfect opportunity and location to start raising a family.[3]

Hillary Rodham was born on October 26, 1947, and three years later her mother gave birth to Hugh Jr., who was followed four years later by Anthony (Tony). Once she gave birth to Hillary, Dorothy became a full-time wife and mother, never working outside the home, and never treating her children or grandchildren the way her mother and grandmother had mistreated her and her siblings. Dorothy showered them with the care and love that had evaded her for much of her adolescence, while Hugh helped to provide a stable and dependable environment for the kids to grow up.

From the start, Hillary seemed born with a strong, determined personality, full of confidence and certitude and tenacity, much like her father. While Dorothy was an influential force among her children, it was Hugh who dominated the family and always made his presence felt within the Rodham household. His parenting was governed by many of the traits that had made him an effective leader on the field and an effective trainer in the navy. He possessed a tenacity and unrelenting competitiveness—a constant drive for perfection that came to have a profound impact on the personalities of his children.

The Gospel According to Hugh

Hugh brought his tough approach to life into his child-rearing practices. The Rodhams, Hugh preached, were self-reliant and self-sufficient. The only help that could be requested was from God alone. To that end, he routinely held himself up as a model of self-reliance that he expected his children to follow. After all, he had managed to put himself through college during the Great Depression; if he didn't need help then, why should anyone?

To Hugh, the purpose of the government was not to be a nanny, and in that respect, Hugh Rodham embodied the Republicanism that rejected FDR and the New Deal. Individuals should not look to the government for a handout; they should look to themselves and to the Lord. It was this thinking that made Hugh a Hoover Republican during the Depression and beyond. "Hugh always voted Republican," said a friend, "and not just voted, but he could be downright righteous and rabid about it."[4]

This doctrine of self-reliance and faith in God was one that Hugh employed in every aspect of his life, and he went to great lengths to instill the same set of core values in his children. For the most part, it was not this order in and of itself that was problematic for his family; it was more that he failed to soften his rigid value system to accommodate the love and warmth that children require. One Hillary biographer, Gail Sheehy, claims: "Pop-Pop, as the children called the authoritarian drillmaster at the head of the family, neither offered nor asked for nurturing. Matters of the heart were a fickle distraction in the Rodham household. Life was seen as combat."[5]

There were practical explanations for Hugh's philosophy, much of which could be traced to his rough-and-tumble youth, having been nurtured in the mines, having decked opponents on the football field, having hitched rides on railroad cars, having belted his way through the Great Depression, having made boys into men in the atmosphere of a boot camp. He emerged from those experiences with his mind and his views on God fully developed, and he was not afraid to let those around him know it. Says Sheehy: "[H]e gave a good imitation of General Patton in raising his children." He would turn to his little girl and ask, "Well, Hillary, how are you going to dig yourself out of this one?"[6]

For her part, Hillary has portrayed her father as not quite so Patton-like, and by her own accounts, including those in her memoir, *Living History*, he is portrayed as a kind, loving father, contrary to the biographical sketches that frame him as a cold, stern Republican taskmaster.

While Hugh is frequently vilified for his domineering role in Hillary's life, in many accounts, Dorothy is often portrayed as completely submissive to Hugh—almost fearful and cowering. But to paint the dynamic as that cut-and-dried is to oversimplify. Dorothy was no shrinking violet. Like Hugh, she had cultivated her own skills of independence during her years when she was forced to rely on herself. Like Hugh, she believed that her children needed to use their own strength and the strength of God to get them through their problems. Like Hugh, Dorothy could have little sympathy if she thought her children were not taking charge of their fates.

When the family moved to Park Ridge, four-year-old Hillary ran into a bully of a little girl named Suzy, a merciless toddler who regularly belted both boys and girls, including the sweet, beribboned Hillary. Each time she walloped Hillary, Suzy exulted in victory as tiny Hillary dashed home crying. Dorothy would have none of this: "There's no room in this house for cowards!" she informed her daughter. "The next time she hits you, I want you to hit her back."[7]

The next time little Hillary was confronted by the brat, who had been encouraged by a pack of mocking, villainous boys there to imbibe in Suzy's cruelty, the Rodham girl shocked everyone by raising her trembling fist and punching Suzy, knocking her off her feet. The boys stood there, mouths agape, as the stunned tyrant fell to the ground. The triumphant Hillary sprinted back to the house and told her mother, "I can play with the boys now!"

It was an important moment for Hillary and one that Dorothy would later come to recognize as crucial to the development of her daughter's character. The altercation with Suzy changed the way Hillary interacted with everyone—especially the boys. Dorothy Rodham said with a measure of great satisfaction: "Boys responded well to Hillary. She just took charge, and they let her." According to Sheehy, Dorothy took pleasure in molding a daughter who compensated for her own "submission to patriarchy," a lesson that sank deep into the little girl's bones.[8] Never again would Hillary submit to the will of men; she would be always ready and eager to mix it up with the boys.

Still, when it came to no nonsense, no one could hold a candle to Hugh. And his son Hughie, like his daughter, had some lessons that Hugh felt needed learning. Hughie followed in his father's footsteps onto the gridiron, achieving a local celebrity for skill—not enough, however, to satisfy Hugh. In one game, Hughie's team stomped its opponent 36 to 0, with the young Rodham leading the way at quarterback, completing ten of eleven passes. The old man felt the compulsion to at least act unimpressed, telling his boy, "You should have completed the other one."

It was this attitude of perfectionism that Hugh forced on all his children. But though he may have been striving for perfection, the end result was often something much different. An early Hillary biographer, the late Barbara Olson, who shared this anecdote about Hughie's football prowess, noted that Hugh's motivation was surely to inspire overachievement in his children; in fact, however, he prompted feelings of longing and never being able to meet a father's fanatical expectations. In Hillary's case, writes Olson, Hugh produced an opposite but positive effect by creating in his daughter a "reservoir of resentment" that may have kindled her to later write an entire book on the value of nurturing.[9]

While Hillary's focus on care and nurturing was an unintended positive of Hugh's regimen, there were many unintended negatives as well that would later play crucial roles as she began her political life. His coldness, emotional detachment, stern demeanor, and lack of charm would all make their own impression on his daughter, eventually becoming the quintessential and routine criticisms of Hillary in her public life.

Hugh and the Park Ridge Church

Though some of Hugh's behavior toward Hillary clearly had some adverse ramifications, not all his contributions were bad, and perhaps the most notable of his roles in her life was how he helped to shape the foundation of her faith. Hugh came from a long line of Methodists. Hillary herself has spoken often and openly of her family's Methodist roots:

Historically my father's family was always Methodist and took it very seriously. Mine is a family who traces our roots back to Bristol, England, to the coal mines and the Wesleys. So as a young child I would hear stories that my grandfather had heard from his parents, who heard them from their parents who were all involved in the great evangelical movement that swept England.[10]

But though Hugh's ancestry steadfastly connected him to Methodism, he did not attend weekly services at the First United Methodist Church in Park Ridge where he and his family worshipped. A congregation with well-established roots in the Park Ridge community, First United is housed in a prominent and wealthy redbrick building, large enough to accommodate the three thousand members of the church. Over the years, Hillary has described First United as a "big church" with a "very active" congregation, located only a few blocks from the Rodham home, so close Hillary and her mother often walked to Sunday services.

Despite the breadth of the congregation, Hugh remained fairly removed, if not seemingly altogether absent, from the parish. A typical assessment of Hugh's involvement in the church, or lack thereof, comes from Leon Osgood, one of young Hillary's Sunday school teachers, a church leader for many years, and a member of First United all his life: "He was not active in the church for many years. . . . Hugh was not active at all. He traveled a lot on business, maybe that was it."[11] Most reports agree that Hugh was "seldom seen" at church.[12]

This is not to say that Hugh was not a believer. Hillary remembers of her family: "We talked with God, walked with God, ate, studied and argued with God. Each night we knelt by our beds to pray."[13] And Hugh enforced the discipline of this faith-driven routine with his typical military precision. Hillary says that it was Hugh who taught his children to kneel by their beds and pray every night. "Our spiritual life as a family was spirited and constant," Hillary later wrote, and Hugh was a central reason.[14]

Much of Hugh's faith seemed inextricably linked to the Methodist upbringing of his youth. He had no special allegiance to the church in Park Ridge, and in fact he was so dedicated to the church of his roots that he and Dorothy drove each of their babies back to the Court Street Methodist Church in Scranton to be christened in front of Hugh's father and relatives. Though Hillary and her brothers were born in Chicago, Hugh ensured they were all baptized near the Pennsylvania coal mines.[15]

If Hugh brought religion into the home, Dorothy made sure that Hillary and her brothers traveled outside to get it as well. Dorothy was much more involved at the Park Ridge church than Hugh, teaching Sunday school and attending weekly services with her daughter.[16] It was left mainly to the Rodham ladies to do the churchgoing. Dorothy later confessed that one reason she taught Sunday school was to keep an eye on Hillary's brothers, to make sure they actually showed up and stayed after they were dropped off.[17]

As one might expect, this lack of fatherly "faith modeling" affected the two Rodham boys. Of her and her siblings, Hillary was the only one who continued to attend Sunday worship services upon reaching adolescence, and much of this was thanks to Dorothy's role in her daughter's faith.[18] In Dorothy, Hillary had something of a spiritual model, a person who showed her how to interact with a community of faith and demonstrate devotion to God through service to others. Dorothy showed Hillary what the public face of faith looked like.

The combination of Hillary's mother teaching Sunday school and her father insisting on looking upward for sustenance and guidance had a pervasive impact on Hillary's spiritual development. In particular, these disparate but overlapping influences of her parents made prayer a lasting part of her life, playing a key role in the fact that, to this day, she sees prayer as fundamental to a lively faith. Hillary is quick to note that from her earliest days, throughout "the daily back and forth of living," especially in rough times, she has found prayer to be "a very important replenisher."[19]

Dorothy and Hugh were not the only people working to shape Hillary's faith; everywhere she turned she seemed to gain new design and guidance over her faith. Her face was a common sight at Park

Ridge Methodist in the 1950s, as she became an integral part of life at the church, and the church, in turn, became an integral part of her life. It became, she said, "a critical part of my growing up . . . I almost couldn't even list all the ways it influenced me, and helped me develop as a person, not only on my own faith journey, but with a sense of obligation to others." She learned from the ministers and lay leaders "about the connection between my personal faith and the obligations I faced as a Christian, both to other individuals and to society."[20]

Hillary remembers the church as "the center" not only for Sunday morning worship but also for Sunday evening groups and other activities during the week. Park Ridge Methodist had a senior pastor, a youth minister, a minister in charge of pastoral visitations, a music director, and a staff member who specialized in Christian education. The congregation built a new wing of classrooms to accommodate the burgeoning number of children ushered in by the postwar baby boom, children like Hillary.

Because of Dorothy's involvement, the Christian education element of the church became a big part of Hillary's life. Not only did Hillary attend Sunday school, but she also went to the annual summertime Vacation Bible School, from which she vividly remembers songs like, "Jesus loves the little children of the world, red and yellow, black and white, they are precious in His sight."[21]

Hillary said that her personal experiences in church as a child were "so positive—not only the youth ministry work that I was part of but a really active, vital, outreaching Sunday school experience, lots of activities for children; there was a sense in which the church was our second home."

Church was more than simply a place for worship; it was a place for life. They went there to study God and read Scripture lessons, but also to help clean up, to play volleyball, to go to potluck dinners, to be in plays, to participate in Christmas and Easter pageants. "It was just a very big part of my life," said Hillary. "And that kind of fellowship was real important to me."[22]

Hillary's Methodism

Though the general concept of faith was important to Hillary during her formative years, she found herself drawn to the specific doctrines of Christianity that were taught in the Methodist church. Her views at that time and still today were highly influenced by, as she put it, that "wonderful old saying" of the church's founder, John Wesley, about doing "all the good you can." Wesley's rule, says Hillary, was: "Do all the good you can, by all the means you can, in all the ways you can, at all the times you can, to all the people you can, as long as you ever can."[23]

To this day, she remains attracted to Methodism's "emphasis on personal salvation combined with active applied Christianity," and how the faith serves as a "practical method of trying to live as a Christian in a difficult and challenging world."[24] Hillary has always been drawn to, in her words, the "approach of a faith . . . based on 'Scripture, tradition, experience, and reason,'" which she describes as the guideposts of Methodism. "As a Christian," she said, "part of my obligation is to take action to alleviate suffering. Explicit recognition of that in the Methodist tradition is one reason I'm comfortable in this church."[25] She believed that the church at Park Ridge stayed true to that mission; it was a "center for preaching and practicing the social gospel, so important to our Methodist traditions."[26]

This Park Ridge definition of Christianity's role in society played a crucial role in establishing in young Hillary's mind what it meant to be a Christian. The social gospel message of the Methodists resonated with Hillary more than any other religious teaching, and the extent to which Hillary's personal faith has mirrored that of the Methodist leadership is remarkable—on both issues and attitude. As the twentieth century plowed ahead, influential elements within the Methodist movement pushed the denomination in the direction of not only progressive thought but also socialist leanings in the realm of economics, at times even edging toward utopianism and a belief in human perfectibility.

But though her adherence to Methodist doctrine as a young woman appeared steadfast, the revolution of the times was beginning to seep into Park Ridge. Changes were encroaching upon the First United community, whether the community was ready or not. These developments would have a profound impact on not just this church, but churches like it across the country, as significant social upheaval tested people's definitions of faith and challenged their deeply held views about the role of God in their lives.

For most people, these changes would come in all different shapes and sizes in the course of the 1960s, but for Hillary, these changes came to town driving a red convertible.

CHAPTER 2

The Don Jones Influence

For much of Hillary's youth, Hugh was the only male influence on her life that had any real bearing. Her views, her political ideas, her religion—all of this was filtered through Hugh, as he helped shape her sense of the world and her sense of self. But all that changed dramatically when she was thirteen, and the Reverend Don Jones, a Methodist minister, entered her life.

Fresh out of divinity school at Drew University, Jones came to Park Ridge to be First United's youth minister.[1] When he arrived, Jones was in his late twenties, and with his blue eyes, blond hair, and red convertible, he proved to be a striking contrast to the three previous youth ministers, all of whom, by comparison, were fossils. As such, no one in the congregation was prepared for a newly minted minister like Jones.

Prior to Jones's arrival, there was no doubt that First United was a conservative congregation. Don Jones hoped to shake it up by guiding the youth group in a totally new way. For Hillary and the other teens, practicing religion had always been a combination of listening to the senior pastor's sermons on Sunday morning and interacting with the youth minister, whose job it was to act as a spiritual guide for the young minds. Starting in September 1961, Jones arranged to have meetings with the youth group on Thursday evenings, and it was there with his "University of Life" program that Hillary fully discovered Don Jones.[2]

When he entered the First United congregation, it was difficult to categorize Jones's theology. Some sources claimed he was a self-described existentialist; others said he was shaped to some degree by the influential theologians Karl Barth and Reinhold Niebuhr, widely read by generations of both left-leaning and right-leaning Christians. While it seemed from the onset that he was liberal both socially and politically, even this proved difficult to state categorically. Jones walked a fine line between rightly awakening the young folks to the vast social changes happening beyond the world of Park Ridge and indoctrinating them to a particular political point of view. Those sympathetic to his perspective would say he enlightened and educated them, whereas those with whom he disagreed would later charge that he brainwashed them.[3]

Given the social climate of the times, Jones was certainly justified in bringing his youth ministry in contact with the broader currents of society. He had come to Park Ridge as the civil rights movement was gathering momentum and breaking out of the inner cities. Everywhere it seemed that a revolutionary perspective on the world was beginning to take form, and he saw it as his duty to usher his young pupils into this new era, to give them the information they would need to succeed with their faith when the world around them was changing. A religion is a worldview, and as such, it must weigh in on the pressing issues of the day if it is to stay relevant.

To that end, Jones was not timid. He introduced his wide-eyed flock not only to the world of Wesley but also to existentialism, abstract art, beat poetry, and even the radical politics of the counter-culture.[4] Gail Sheehy maintains that Jones was not as fired by theology as he was by the "cultural revolution" that he felt was under way. Jones asked one of Hillary's classmates, Bob Berg, to play for the youth group Bob Dylan's "A Hard Rain's A-Gonna Fall," as Jones's impressionable disciples ruminated upon the lyrics. He distributed sheets of poetry and handouts in which D. H. Lawrence refuted Plato. He rented a projector to show François Truffaut's classic *400 Blows*, Rod Serling's *Requiem for a Heavyweight*, and other art house films of the time.[5]

Clearly the freethinking, freewheeling Jones was not afraid to make waves. The point was to make real "the feelings of others," as one student remembered, and to enliven the "practical conscience and content" of their faith.[6] If that was the goal, Jones certainly achieved it. One of the youths, Rick Ricketts, whom Hillary had known since she was eight years old, recalled the lively discussions Jones generated: "I remember when he brought an atheist to the group for a debate with a Christian over the existence of God," said Ricketts of Jones's class. "There was also a discussion of teenage pregnancy, which got the whole congregation upset."[7]

Black and White

While many of Jones's teachings were designed to increase the students' general sense of cultural perspective and awareness, he put a lot of emphasis on the church's role within the racial struggles of the time. Racial awareness and activism were integral to Jones's aim of raising social consciousness among the adolescents, and it was a point that Hillary responded to with a great deal of enthusiasm.

"In Park Ridge then," said Jones, "you wouldn't know there were black people in the world."[8] And so, Jones made it his goal to give them firsthand experiences with the larger world. It was not enough to learn about things from books, films, and poetry; Jones wanted his students to live these teachings, to see their faith interacting with the world around them.

He drove the youngsters to a community center on the Southside, where he brought them together with inner-city youth and assembled them all around a print of Picasso's *Guernica* that Jones had brought along and propped up against a chair. The painting graphically depicts a bombing raid by pro-Franco forces on a village during the Spanish Civil War. The minister opened an edgy dialogue over war and violence, asking questions like, "What strikes you about this?" "Any imagery?" "If you had to title this painting with a current piece of music, what would it be?" "Have you ever experienced anything like this?"[9]

As Hillary and her friends spoke of the horrors of war in the abstract, the inner-city kids related the mayhem to their everyday lives. One girl looked at the painting and, to the shock of the Park Ridgers, chimed in, "Just last week, my uncle drove up and parked on the street and some guy came up to him and said you can't park there, that's my parking place, and my uncle resisted him and the guy pulled out a gun and shot him."[10] The girl's response jolted the white-bread kids.

Jones's goal was accomplished. And whether his pupils, or even he, knew it or fully grasped it, he was preparing them for the social upheaval about to befall them and America before the decade ended.[11]

Amid the radical dialogue and rhetoric of Jones, there was another influential young adult who was battling for the heart, mind, and soul of Hillary Rodham and her peers—Hillary's teacher Paul Carlson, a conservative Republican and staunch anti-Communist. Carlson remembers returning to Park Ridge after graduate school and going to the Methodist church to hear Jones speak. He was alarmed by what he heard. "I approached him after the service and suggested we get together. I was teaching at Maine East [High School] at the time and was concerned. I had lunch with him the next day and explained that I wasn't at all amused with his message, that it was not a Christian speech. He really gave me no defense of what he had said."[12]

Referring to the trip to the Southside of Chicago, Carlson complained that Jones's intention was to take Hillary and her white friends to the slums to blame them and their class for the conditions of the inner city and to fill them with white guilt, an accusation that to this day Jones adamantly denies.

Whatever Jones's intention, the encounter had a profound effect on Hillary, as did the moment in April 1962 when Jones took the teenage Hillary and her class back to Chicago to hear a rousing speech by a man named Martin Luther King Jr. at Chicago's Orchestra Hall. There the civil rights pioneer preached a sermon titled "Sleeping Through the Revolution," and the experience gave Jones the opportunity to leave yet another indelible mark on his pupils. "I wanted them to become aware of the social revolution that was taking place," says Jones of his Park Ridgers. "Here was the major leader of that movement, a Protestant preacher, coming to town. It was an opportunity for them to meet a great person. . . . Park Ridge was sleeping through the biggest social revolution this country has ever had."[13]

Dr. King could not have agreed more. In his speech, King said that too many Americans were like Rip Van Winkle, snoozing through the historic changes happening all around them.[14]

That night was one Hillary would never forget, particularly because of the moment after the speech, when Jones shocked the teen and her comrades by arranging to have them briefly meet with and shake the hand of King. Jones introduced King to each of the kids, one by one, and Hillary would be forever grateful.[15] Later in life, Hillary would remark that these experiences in Chicago opened her eyes "as a teenager to other people and the way they live [which] certainly affected me."[16]

Church Politics

Martin Luther King Jr. might have been the best-known activist to whom Don Jones introduced his youth group, but there were many others to whom he exposed his class of young minds. Chief on that list was the legendary radical and social activist Saul Alinsky, whom Jones took his group to meet with in Chicago. Born in 1909, the often profane, crude, and always irreverent Chicagoan was dedicated to ripping down the "power structure" throughout capitalist America, and he devoted much of his life to organizing demonstrations throughout the country. For much of Alinsky's life, rumors of his affiliation with the Communist Party surrounded his actions, and his skilled rhetoric often had a strong appeal to many people on the left.[17] Alinsky penned *Reveille for Radicals*, the 1946 bible of the protest movement, forever establishing him as "the father of community organizing." The organizer considered himself a survivor of the "Joe McCarthy holocaust of the early 1950s," who was now at the "vanguard" of the "revolutionary force."[18]

Here, too, Jones left a lasting impression, and Hillary would later describe Alinsky as a "great seducer" of young minds, but then again, that may have been the very reason Jones brought the group to Alinsky in the first place. In truth, Jones's goal in introducing his acolytes to Alinsky could not have been all that religious, since Alinsky was a well-known and committed agnostic Jew who proudly declared his "independence" from any affiliation, including Christianity. Hillary was among those taken in, so intrigued and impressed by Alinsky that she would later write her college thesis on his strategies.[19]

It certainly would not have been out of bounds for Park Ridge Methodist parents to question the relevance of such introductions to the church youth ministry, since it was becoming increasingly clear that the trips, like the one to visit Alinsky, were simply another vessel through which Jones could advance his progressive message. But though some parents questioned many of Jones's decisions, they continued to support First United's choice of Jones as youth minister.

Personal Jones

As his relationship with the group developed, it was becoming clear that Jones was advancing his own brand of socially conscious Methodist thought. Using theologian Paul Tillich as his spiritual mentor, Jones believed the most important role of the church was to help the less fortunate. He drew the social justice lines in bright colors on the chalkboard inside the Methodist church, and with his persuasive arguments, he began to win over Hillary and some of her classmates. Said Hillary, "He was just relentless in telling us that to be a Christian did not just mean you were concerned about your own personal salvation."[20] Instead, he preached a message that to be a Christian meant that one also had to be concerned about others—particularly those with less money.

In the spirit of Alinsky, by late 1963, Jones had his pupils busy organizing. They set up food drives for the poor and even coordinated a ministry to the children of migrant farm workers, with Hillary taking a lead role in the latter. These Hispanic laborers were trucked in for temporary farming work. They lived in squalor amid fields west of Chicago—not far from Park Ridge. Fourteen-year-old Hillary organized groups of older students into a babysitting pool, while she served cold drinks and cupcakes. Some of the girls sewed children's clothing and doll clothes.[21] Meanwhile, Jones taxied Hillary and the other teen girls to the migrant camp in his convertible.[22]

Hillary responded positively to Jones's message of social responsibility, and in Hillary, Jones saw the potential for really activating a teen and molding her. The relationship between Jones and Hillary was intellectual, spiritual; still, in that way, it was also personal. Hillary was hooked by Jones, and he was impressed by her. She immediately struck him as a "straitlaced Methodist," but the more they worked together, the more he discovered that there were other layers to her. "She was serious, but she was also gregarious," Jones said of Hillary. "She wasn't the cheerleader type, but she wasn't the shy bookworm, either."[23]

At twice her age, Jones obviously viewed himself as purely a mentor to her, and Hillary, who was not the giddy, girly, boy-crazy type, was far more interested in the brain of an adult like Jones than the looks of a boy from math class. She began dropping by Jones's office after school or during summer afternoons, eager to talk about ideas or insights she culled from the youth minister and his sermons. According to Roger Morris, biographer of the Clintons, Jones had her read Tillich, Niebuhr, Søren Kierkegaard, and Dietrich Bonhoeffer, and they had lengthy, increasingly serious discussions. "She was curious," says Jones. "She was just insatiable."[24]

They discussed Tillich's reformism, Kierkegaard's "leap of faith" in the face of rational cynicism, Bonhoeffer's "religionless Christianity," and especially Niebuhr's views on history, human nature, and the necessary force of civil governance—all of which played crucial and formative roles in her constantly evolving sense of faith. "She realizes absolutely the truth of the human condition," Jones took from Hillary's study of his good works, adding this important insight: "She is very much the sort of Christian who understands that the use of power to achieve social good is legitimate."[25]

Morris says that rather than weighty discussions among intellectual equals, these talks between Hillary and Jones were more akin to "tentative discoveries . . . the first fitful awakenings of critical intellect and sensibility in a spiritually minded young woman." He says that Jones estimated that Hillary was, at heart, a cautious, contained, "self-protective girl" whose judgments about herself and her world were still forming, and as such she required constant intellectual direction. When he gave her a copy of J. D. Salinger's *Catcher in the Rye*, Hillary originally balked at the book, feeling that it was too heavy for her to understand.[26] "I thought it was a little too advanced for me," she later wrote to Jones during her sophomore year in college, when she was assigned the book in an English class and was able to better comprehend it the second time around.[27] His reading list offered spiritual comfort as well, as his students did discuss the Bible, and, apart from Salinger and Niebuhr, Jones was prone to leave pupils like Hillary with small Methodist devotionals—books with inspiring Methodist interpretations of Scripture—to carry for sustenance.

Morris adds that Jones was not only intellectually exciting to Hillary but nurturing, approving, accepting, and embracing. He was the "world beyond" the "growling Hugh Rodham."[28] By this point, Hillary was conflicted, stuck in a political purgatory between the politics of Don Jones and those of her father. "I wonder if it's possible to be a mental conservative and a heart liberal?" she wrote at the time, a kind of early tug toward what George W. Bush later coined "compassionate conservatism"—a phrase that probably would have had tremendous appeal to her at the time.[29]

Overall, Jones must have been pleased by this progress in his star pupil. In Jones's mind, these were exactly the types of questions he wanted to hear from Hillary, and he understood that this "deeply religious" girl was still growing into herself. She was "so nonfrivolous it's unbelievable," he said. "Hillary was curious and wide open to everything—not that she liked everything and accepted everything at face value."[30] A third party, Rick Ricketts, a fellow student of the University of Life sessions, remembers how Hillary became Jones's chosen disciple: "She seemed to be on a quest for transcendence," said Ricketts. Even after they went their separate ways, Hillary would continue to see Jones as a mentor, evident in the long, earnest, sometimes painful letters she wrote to him during high school and college, which were often filled with celestial navigations about life and philosophy and contemplation of her quest to express her faith through social action.[31]

Jones Moves On

Eventually, Jones's waves proved to be too big for the congregation. Disturbed by him and his activities, they forced him to leave the Park Ridge community two or three years short of the typical stay for a youth minister. Though he was popular with the youth, the majority of the adults viewed his tenure as harmful.[32] Authors Peter Flaherty and Timothy Flaherty quote Bob Williams, a lifelong Park Ridge resident and church member, who remembered Jones's stint as "extremely disruptive." He said people of the town sarcastically referred to Jones as the "Marvel from North Dakota." "He looked like an All-American, blond hair and tanned," said Williams. "He used to drive around town in a convertible—the girls salivated."

Williams was "amazed" when he learned years later that First Lady Hillary Rodham Clinton considered Jones one of her mentors: "I understand that Don went on and did some other things, but from what I knew of him then, if you were picking mentors—he's one you would not pick. People were leaving the church because of his teachings."[33]

Flaherty and Flaherty also quote Leon Osgood, who knew Jones well:

I liked Don and considered him a good friend. I felt sorry for him then; I thought he was at least well meaning. But he got under the skin of the older members of the church and they forced him to leave. He was young and show-offy, driving around in his convertible. We had a large youth group and the feeling was that he was having an influence on quite a few people. . . . I actually agreed with him on some of his teachings. But, I was one of the few. His new ideas on sexuality and things like that were not welcome in the church. . . . The older members were quite concerned, and asked him to leave.[34]

Osgood wrote a recommendation for Jones after he was pushed out. He felt it was "the least I could do for the guy. . . . I was glad to see him get that other job."[35]

In 1964, Jones was assigned to another church after little more than four years in Park Ridge, and before Hillary finished high school. Now Hillary had a new youth minister, one who faced a new Hillary, the Don Jones version, quite different from the Hillary of four years earlier. She wrote of the new minister in a letter to Jones: "He thinks I'm a radical."[36] Jones must have grinned.

While Jones's stay in Park Ridge had been brief, it had been quite influential. It would be unfair and disrespectful to suggest, as some have, that Jones was a "guru" to Hillary Rodham.[37] After all, he did not instruct her in trendy New Age ideas, nor preach some New School of Social Thought sophistry. Rather, he was, plainly, a very influential person in her life, who—to the extent that a sympathetic, religious-minded observer would search for signs of the hand of Providence—seemed to be placed in her life at just the right time. The fact that Jones came to her during those critical high school years, those years of teen angst and development, proved immeasurably timely, since if he had come just two years later, he might have encountered a more uphill battle when dealing with Hillary.

In the end, Jones's emphasis on social duty as the bedrock of faith provided a spark in Hillary's budding relationship with God, successfully eroding some of Hugh Rodham's most influential and deeply rooted teachings about responsibility to oneself and the world at large. But although she was strongly affected by Jones and moved by his litany of causes and ideas, it was still too soon for a complete revolution. Instead, it was time for a different sort of change. It was time to go east.

CHAPTER 3

Hillary Hits Wellesley

Despite his influence, Don Jones could not in four years completely dislodge from Hillary a decade and a half of Hugh Rodham's Republicanism. In the fall of 1964, the high school senior soaked up Barry Goldwater's *Conscience of a Conservative* and devoted her term paper to the then-fledgling American conservative movement, which, after an initial boost with the founding of a young William F. Buckley Jr.'s *National Review* magazine in 1955, now had its first ideologically pure presidential candidate in Goldwater. Already an active member of the local chapter of the Young Republicans, she signed on to campaign for the Arizona senator and Republican presidential candidate as a "Goldwater Girl." To the delight of her father, she went all over, canvassing solidly Republican neighborhoods, organizing classmates for Republican events, and even, as she would sheepishly admit later, donning the corny cowgirl getup and hat with the symbol AuH_2O.[1]

But this devotion to the cause of conservative Republicanism was fleeting. On that, a number of Hillary's high school friends state that the supposed political-philosophical sea change about to befall

Hillary Rodham has been exaggerated. Her friends say that the early Goldwater support and her Park Ridge Young Republicanism was merely the prototype of a young person following her parents' example, and that Hillary was never an actual conservative.[2]

This interpretation appears at least partly correct. She thought about the issues, but did so without a lot of interaction or testing from the outside. Hers was never a tried-and-true conservatism; very few sixteen-year-olds have had the experiences and time to thoroughly think through the political philosophy that they will carry with them the rest of their lives.

"I don't think that Hillary was ever really conservative except in a fiscal sense," says Dorothy Rodham. "Her father was extremely conservative in that area." Even as a fiscally conservative Republican, says Dorothy, Hillary was driven by a desire for social justice.[3]

She was disappointed when Goldwater lost, but she quickly turned her attention to another horizon: college. She applied to Radcliffe, Vassar, and Smith, and according to her mother, Hillary was admitted to all of them.[4] She was attracted to the merits of single-sex education. She opted for Wellesley College in part because one of her favorite high school teachers had gone there. In June 1965, Hillary left high school, where she had excelled academically. At graduation, she was called to the stage several times to receive various academic awards, from National Merit Scholar finalist to National Honor Society. Dorothy and even Hugh couldn't help but blush.[5]

1965–1969: The Wellesley Girl

In the fall of 1965, Hillary Rodham left Park Ridge for Wellesley in Massachusetts, where she enrolled with roughly four hundred other freshmen at the all-women's school. The good little Goldwater Girl was forewarned by her high school government teacher, Gerald Baker: "You're going to go to Wellesley, and you're going to become a liberal and a Democrat." Hillary objected: "I'm smart. I know where I stand on the issues. And I'm not going to change."[6]

Though colleges of the 1950s and early 1960s were traditional, staid, tweedy, laden with rules and curfews, this did not mean that the classrooms were politically conservative. On the contrary, Wellesley, like many other centers of higher learning, found itself to be at the center of the cultural revolution that Jones had spoken of so frequently. The faculties at these colleges were by and large liberal, and the occasional professor who was a Republican was a Rockefeller Republican, as were many of the parents who sent their boys and girls to these colleges—with the exception of Hugh Rodham.

While the Wellesley of Hillary's time was relatively open to the practice of religion, it was hardly a conservative Christian college. And like many God-fearing Republican parents, Hugh Rodham naively handed over his Goldwater Girl to a faculty whose priorities were not in line with those that Hugh had outlined during the formative years of Hillary's life. At Wellesley, the changing times were prominently on display, and it was a rolling tide that Hugh could do nothing to halt. In this milieu, after four years of Don Jones, the stage was set to complete Hillary's transformation from a street-canvassing Republican who believed entirely in her father's doctrine of self-reliance and faith in the Lord, to the woman who would one day seek to socialize American health care.

Wellesley in 1965 was a remarkably homogeneous place, with its population essentially at 99 percent white female. There were only six African American students in Hillary's class, whom many of her classmates referred to as "Negroes," and there were no black faculty members.[7]

In this unbalanced setting, Hillary immediately sought out the black students and continued her interest in civil rights that she had cultivated under Don Jones's guidance. Not long after her arrival at Wellesley, Hillary began attending services at a nearby Methodist church, but it did not take long for her unique brand of Jones's faith to make an appearance within her church's walls. About a month into her first semester, Hillary invited a black classmate to attend Sunday church services with her at the Methodist church, a move that raised eyebrows—even in a place as progressive as Massachusetts. Afterward, she telephoned some Park Ridgers to boast of her bold stroke; some of them were alarmed, and expressed misgivings. She wrote to Don Jones about the experience, including the reaction at their old church. Jones later recalled that the Park Ridge Methodist folks were bothered because Hillary seemed to make the move "not out of goodwill" but simply to shock a "lily-white church."[8] To this, she happily pleaded guilty, admitting to Jones that she was "testing the church."[9]

This is not to say that her gesture was not heartfelt. She told Jones that she was testing herself as well. She was genuinely interested in her minority classmates, and today, African American schoolmates like Karen Williamson speak warmly of Hillary and how she treated them: "She was just a friend. And as a black woman going to Wellesley at the time friends were very welcome. She was warm, a regular person. All the black students in our class felt we had a very close friendship in Hillary." They also sensed something more: "A lot of us thought Hillary would be the first woman president," said Williamson later. "I thought if ever in my lifetime there is a woman president it would be her."[10]

Hillary's interest in civil rights was not limited to the Wellesley campus. Despite the rigors of a packed college schedule, Hillary found the time to volunteer to teach reading to underprivileged black children in Boston's Roxbury neighborhood, continuing on and pursuing her personal ministry of social duty.[11] She was loving her neighbor as herself, and by doing so she was staying true to her Methodist faith.

From her first days on campus, she ran for office, a fact that prompted observers and later biographers to be suspicious of her motives when she did and said nice or popular things—a trait even more prominent in the decades ahead. Her friends at Wellesley talk of how "recognition was important to her."[12] She had been elected president of the Wellesley Young Republicans in her freshman year. By her sophomore year, however, she had begun rapidly moving left. To cite just one example, she was thrilled with the election of liberal Republican John V. Lindsay as mayor of New York. She shared her excitement over Lindsay in a letter to Don Jones, underscoring how she was "leaning left"—"See how liberal I'm becoming!"[13]

Despite the left-leaning advances that she seemed to be making on her own, it was Martin Luther King Jr.'s assassination that produced one of the most dramatic shifts in her college behavior. One April day in 1968, Hillary's roommate, Johanna Branson, was in her suite when the door was suddenly thrown open and a distraught Hillary threw her green book bag across the room. She yelled, "I can't stand it anymore. I can't take it!" Hillary quickly called Karen Williamson, who headed the black student organization, to express her sorrow.[14]

Hillary was angry—very angry. The trauma of that bullet seemed to catalyze Hillary's politics. Nevertheless, her classmates insist she was never a "radical," perhaps because unlike more militant activists of the time, Hillary was more willing to work within the system to change things, using the proper channels and procedures, and putting together coalitions through student government, rather than tearing down the administration or calling for the dean's and provost's resignations.[15]

Only weeks after the King assassination, however, Hillary helped lead students in a two-day strike on campus—a "teach-in" in which students did not attend class for two days and instead participated in seminars on race and the Vietnam War. When Professor Marshall Goldman, the famous Russia scholar and one of the best-known faculty members at Wellesley, suggested to Hillary and an assembly of students that they make a real sacrifice and give up their weekend for the teach-in, Hillary snapped back: "I'll give up my date Saturday night, Mr. Goldman, but I don't think that's the point. Individual consciences are fine, but individual consciences have to be made manifest." For some reason, emotional students erupted in applause at this unclear statement, perhaps impressed by her use of words like "consciences" and "manifest."[16]

It was incidents like this that probably prompted Ruth Adams, the Wellesley president, to reflect that Hillary was "not always" a pleasure to deal with if you disagreed with her. "She could be very insistent," judged Adams. Agreed a classmate, Gale Lyon-Rosenberger, "She could be a little cutting." Lyon-Rosenberger said that in class Hillary was often challenging, saying of her tone: "It sometimes had an edge to it."[17]

Liberal Motive

Though politics and current events continued to push Hillary away from her father's Republican brand of the Methodist faith, Hillary continued to attend church during her undergraduate years, even joining the interdenominational "chapel society." In addition, she took a year of required Bible study, a class that was mandated by the Wellesley of the day, but that proved to be an experience for which Hillary was "very grateful."[18]

Throughout her time at Wellesley, Hillary kept in close contact with Don Jones, who continued to act as a spiritual and political mentor, turning her on to a now-defunct magazine for college-aged Methodists called *Motive*. Jones had given her a subscription to the magazine as a high school graduation gift, and Hillary subsequently devoured every issue.[19] ("I still have every issue they sent me," she said in a 1994 interview.) *Motive* was not the typical, slightly left of center, Christian publication aimed at college kids. Instead it was an extremely progressive journal, founded in 1941 by Harold Ehrensperger, a pacifist committed to the social gospel tradition and a member of the far-left National Council of Churches. Throughout the 1950s and into the 1960s, *Motive* crept further and further to the left, eventually becoming a magazine for counterculture Christians, whose pages were filled with a kind of who's who of radical Protestants.[20]

The defining shift in *Motive*'s push to the left came in the late 1960s, when the magazine was under the leadership of its fourth editor, B. J. Stiles. During this period, the magazine's special targets were the Vietnam War, anti-Communism, and American "economic imperialism," especially alleged U.S. evildoings in Latin America.[21] Hillary particularly recalls an article by Methodist theologian Carl Oglesby, called "Change or Containment," which had the important effect of pushing her further against U.S. policy in Southeast Asia. "It was the first thing I had ever read that challenged the Vietnam War," she says.[22] Oglesby's article was based on a book by the same name, which he coauthored with Richard Shaull in 1967. Oglesby's words in the book were a scathing blame-America-first indictment of U.S. foreign policy, at times thoughtful, informed, even eloquent, but more often rambling and filled with moral equivalency, failing to differentiate essential differences between American and Soviet goals in the Cold War. And, of course, it was dismissive of the seriousness of the Communist threat.[23] The fact that Hillary was impressed by Oglesby's work is powerful evidence of how far to the left the former Goldwater Girl had moved.

Indeed, Oglesby was no run-of-the-mill Methodist pacifist. He was an early founder and eventual president of the radical SDS—Students for a Democratic Society. In the pages of *Motive*, Oglesby asked questions like, "What would be so wrong about a Vietnam run by Ho Chi Minh, a Cuba by Castro?"[24] Oglesby also wrote for the radical magazine *Ramparts*, founded by then-Communists (now conservatives) David Horowitz and Peter Collier, as well as far-left magazines like *The Nation*.

Motive magazine had drifted so far to the left, politically and religiously, that the left-leaning United Methodist leadership questioned *Motive*'s motives, and would eventually strip it of funding. In other words, the *very years* that Hillary was most influenced by the magazine—so impressed that she still owns every copy—the liberal Methodist leadership found *Motive* so irresponsible that its funding was cut. .[25]

Indeed, the final gasps of the magazine took place from 1969 to 1972. An infamous edition was the "women's liberation issue" of early 1971, where the lead piece began, "Here she is, Miss America. Take her off the stage and f– her."[26] By this point in the life of the magazine, readers should not have been shocked: The often-profane Christian periodical had featured a photo of a pretty coed with an LSD tablet on her tongue, run a mock obituary of God, presented a birthday card to Ho Chi Minh, and regularly provided advice on draft dodging and desertion.[27] Nonetheless, the women's liberation issue was too much and prompted the elders of the church to inform *Motive* that its days were numbered. Half the subscribers canceled, but Hillary was not one of them. This loss of funds was a substantial setback for the magazine that tarnished much of *Motive*'s success during its early years within the Methodist church, such as fighting Southern Democrats on racial segregation.

In 1972, *Motive* made its exit with a special theme issue on lesbianism and feminism and the gay-male sexual experience, the former theme being demanded by the lesbian Methodists on staff, who insisted on equal time in the name of women's equality. The issue was so offensive that it was pulled from the press by executives in the Methodist Church's Board of Education.[28]

Vietnam

By the late 1960s, *Motive* magazine was not the only influential force working to encourage Hillary's budding liberalism; the Wellesley faculty also began playing a role that further hardened her against Republicanism and the ongoing war in Vietnam. As a sophomore,

Hillary worked as a researcher/babysitter for a Wellesley professor who was writing a book on the Vietnam War.[29] The book and interaction with the professor must have made a difference, as it was during this time that Hillary became an increasingly vociferous opponent of the conflict and was on her way to permanent leave from the GOP. After a summer with the professor, she returned to Wellesley in the fall of 1967 and began campaigning for the liberal antiwar candidate Democrat Eugene McCarthy.

Not surprisingly, Hugh supported the war, and the two had a few sharp discussions on the subject that caused some rancorous dinnertable talk. Hillary's abandonment of the Republican Party caused distress to Hugh, who was now reaping the consequences of sending his daughter to a liberal college.

Still, the fact was that despite all her turns and the continued intertwining of her spiritual and her political life, she had not completely forsaken the GOP, so long as the Republicans put forth a more liberal candidate. When Eugene McCarthy left the race, Ms. Rodham volunteered for Nelson Rockefeller, the preference of many liberal Republicans as the best remaining antiwar candidate once McCarthy was gone. She went to the Republican convention in Miami to support Rockefeller. Hillary could tolerate a Rockefeller or a Lindsay, who were anathema to conservatives, but would not dare favor Richard Nixon.

That summer of 1968, she interned as a researcher for the House Republican Conference, chaired by Melvin Laird, who would become President Nixon's secretary of defense. She was assigned to the Republican House member from Park Ridge, meaning that Hillary spent her summer studying the dismal economics of revenue sharing, and oddly enough, her intern supervisor was a young man named Edwin Feulner, who soon launched the enormously successful conservative think tank, the Heritage Foundation.[30]

In the end, that summer was the final thread holding her to Hugh's party. The more she was pulled away from conservative Republicanism, the more she connected with secular liberalism. As the sixties had progressed, Don Jones ceased to play the decisive role in her life that he once had, and without Don Jones as a rudder, she was beginning to find herself in the lonely camp of many liberal Christians—among a sea of secular Democrats who searched for salvation in politics rather than religion.

A serious boyfriend of hers from Harvard, Geoffrey Shields, whom she dated for two years, speaks of Hillary's intense interest in politics during this time, discussing issues up and down, "lighting up" when certain topics were raised, and especially driven by race, civil liberties, and the war.[31] Politics dominated her interests, increasingly at the expense of her Christianity, though Shields recalls some philosophical conversation, such as debate over "whether there was an absolute or only a relative morality."[32] Her answer, however, was not that of a fundamentalist: "I believe there are some absolute truths," she wrote, absolutely, to a friend named John Peavoy. She said that she was unsure as to whether such truths were within man's grasp, though she was certain they existed, and suggested to Peavoy that he read John Stuart Mill's *On Liberty*.[33]

At the same time, Hillary's faith—or perhaps her personality or seriousness generally—must have been a contributing factor to her staying on the straight and narrow. She called herself "an ethical Christian," physically aloof from the counterculture. Her college friends do not recall her smoking dope, dropping acid, drinking to excess, or tearing off her clothes during rock concerts. She did not imbibe the hedonism and drug culture of the period; she did not drop out.[34] She at one time painted a flower on her arm and wore tie-dye clothes, and as surviving photos attest, looked like a girl of the sixties, but was no Janis Joplin.[35]

By 1969, her senior year, Hillary's evolution from Republican to Democrat was complete. She was president of the graduating class, a fact soon made public around the country, as she prepared for a memorable good-bye at Wellesley that May. The commencement speaker that year was Senator Edward W. Brooke (R-Mass.), who in 1966 became the first African American elected to the U.S. Senate since Reconstruction—later awarded the Presidential Medal of Freedom by President George W. Bush. Brooke came to extend his congratulations to the 401 women, but did so to the dissatisfaction of Hillary Rodham, who judged that the good senator had missed the paramount issues driving the times. That was an opinion she did not keep to herself, as the Wellesley brass soon learned to their horror.

The powers-that-be at the college had decided that this year would be the first in which a graduating senior was permitted to speak at commencement. Hillary ensured that the administration would regret its decision.

Though she had spent weeks preparing her text, Hillary tossed aside the script as she approached the platform. She then launched into a stern point-by-point rebuttal of the senator's remarks, with all the moral certainty and righteousness of a preacher. "We feel that for too long our leaders have used politics as the art of the possible," lectured Ms. Rodham. "And the challenge now is to practice politics as the art of making what appears to be impossible, possible." She spoke of her and her sisters' struggles with an "inauthentic reality," a "prevailing acquisitive and competitive corporate life," and yearning for "a more penetrating . . . existence."

She continued her stern—it has been called "scathing"—rebuttal of Senator Brooke, one that got national press, with an excerpt published in *Life* magazine and a front-page article in the *Boston Globe* the next day, the latter of which reported that she had "upstaged" Brooke.

While many of the parents were appalled, Hillary's band of sisters was thrilled. Despite her often clumsy use of ten-cent words and meandering messages, they responded with a standing ovation, or, as more than one source has put it, "enveloping her in a thunderous, seven-minute standing ovation."[36]

To Hillary, the roar of the crowd signaled approval of much more than a brash response to authority; this was a send-off, a commencement all right, a beginning of grander things from a honed Hillary Rodham, ready to take her slice out of history.

CHAPTER 4

God and Woman (and Bill) at Yale

After graduation from Wellesley, Hillary had a variety of opportunities in front of her, but it remained to be seen whether she would pursue higher education or enter the rapidly changing workforce. It was during this ambiguous time that Hillary became reacquainted with an old friend by the name of Saul Alinsky.

In the years since Don Jones had introduced Hillary to Alinsky, the veteran radical had relocated to Carmel, California, and was working in a predominantly African-American neighborhood in Oakland. It was there, in his new habitat, that he invited Hillary to come work for him as an organizer during the spring of 1969. The two began a correspondence, but it quickly became clear that though Hillary had closely studied Alinsky's organizing tactics during college, she remained conflicted, if not torn, over joining his cause.

At the time that Hillary and Alinsky were writing to each other, Alinsky was in the process of putting the final touches on his crowning work, *Rules for Radicals*, which would be published in 1971, and which he was committing to paper and a lecture or two as he was corresponding with Hillary. In the book, Alinsky laid out his "revolutionary" ideas, including the development of a new "Trinity" based on class distinctions, by which he perceived mankind as divided into three groups: "the Haves, the Have-Nots, and the Have-a-Little, Want Mores."[1] It is not clear if Hillary received any spiritual direction from Alinsky during this time, but given the extreme nature of his beliefs, it seems doubtful. While Alinsky may have served as a mentor and an example for Hillary in some ways, his beliefs were most likely—or at least hopefully—too much for the Park Ridge Methodist. Consider the feature quote—literally Alinsky's first words—at the start of *Rules for Radicals*: "Lest we forget at least an over-the-shoulder acknowledgment to the very first radical: from all our legends, mythology, and history (and who is to know where mythology leaves off and history begins—or which is which), the first radical known to man who rebelled against the establishment and did it so effectively that he at least won his own kingdom—Lucifer."[2]

Alinsky actually commenced his masterpiece with an acknowledgment to Satan. Nonetheless, Hillary admired Alinsky as much as ever and did take his job offer seriously. Ultimately she turned him down and decided instead to attend law school at Yale. "Well, that's no way to change anything," said Alinsky of the young lady's bourgeois notion of higher education. She replied, "Well, I see it differently than you." She told Alinsky that she saw a "real opportunity" at Yale.[3]

The levelheaded young lady sensed there was something at Yale that would change her life—that would help her make a difference someday. It turned out that there were many things waiting for her there, not to mention a certain someone.

When Wellesley's class president arrived at Yale Law in the fall of 1969, she continued to study as diligently as ever, and eventually settled in. One day while studying in the library, she spotted a rubeish but oddly charming Southern boy looking her way. Never lacking for self-esteem, Ms. Rodham stepped over to the young man and said,

"If you're going to keep looking at me and I'm going to keep looking back, we at least ought to know each other." He blushed. In no time, they were dating.

The Boy from Hot Springs

William Jefferson Blythe III was born on August 19, 1946, in Hope, Arkansas. He was certainly his mother's son. His mother, Virginia, was the daughter of the town iceman and a nurse. After high school, she had followed in her own mother's footsteps and studied nursing in Shreveport, Louisiana, where she met William Jefferson Blythe Jr., whom she wed in a civil ceremony on September 3, 1943, just before he shipped off for war duty in Italy. Upon completion of her nurse training, she moved back to Hope, where she dreamed about building a home with her husband upon his return and raising a child.

But it was not meant to be; only two of the three had a future together. On a summer night in 1946, twenty-seven-year-old traveling salesman Billie Blythe was driving his Buick along Highway 60 outside of Sikeston, Missouri, heading back to Hope from Chicago. As his pregnant wife awaited him, Billie, for whatever reason, lost control and crashed headfirst into a drainage ditch, where he was knocked unconscious. Other drivers saw the accident and searched for the body but could not find it in the dark of night. That was a tragic shame, because Blythe had landed in shallow water, where he drowned. Had there been enough light to find him, he likely would have been rescued and survived.

The son, Bill, that he was leaving behind, who at that point was still in the comfort of his mother's womb, was the third child sired by William Jefferson Blythe Jr., who was notorious for riding around the South and cajoling women to hop into bed with him. His marriage to

Virginia was his third in eight years; the first was to another woman named Virginia when he was eighteen and then to his first wife's sister Faye in 1940.

Bill Blythe Jr.'s death left little Bill to be raised by his twenty-three-year-old mother, Virginia, in a severely dysfunctional town and unhealthy home. Shortly after Bill's birth, his mother moved with him to Hot Springs, a small Southern version of Las Vegas, minus any glitz or glamour. Hot Springs was a sleazy place where prostitution, gambling, and vice ruled; to say it was a poor place to raise a child is to state the obvious. Conditions for poor Bill were deplorable, and unfortunately, they would only get worse.

In 1950, a few years after Bill had moved to Hot Springs with his mother, Virginia remarried, choosing forty-two-year-old Roger Clinton, a bootlegging car salesman, an alcoholic, and an abuser, as her groom. Much of Bill's early memories were shaped by interactions he had with his abusive stepfather. Biographer Roger Morris maintains that not only did Roger Clinton slap and punch his wife but that the abused wife could never bring herself to stop Roger from attacking her little boys. According to Morris, friends and medical sources in Hot Springs recalled that Billy and his little brother, Roger Jr., the biological son to Virginia and Roger, were themselves "beaten and brutalized far worse than anyone later admitted." Morris says that the boys more than once had to go to their doctor's office and even the hospital for injuries. "A member of my family doctored them for some pretty bad stuff—stitches and all that," a Hot Springs attorney told Morris, who adds that Virginia ensured that no records remained to embarrass them later.[4]

A later political friend of Bill remarked, "Now that I know what was really happening inside that [house], I'm blown away. . . . He covered up like a dog burying a bone real deep."[5] Bill Clinton learned from a very young age to compartmentalize his traumas. Biographers have noted that his mom taught him to simply take those troubles and put them in an imaginary "box," locking them up with a key and setting them aside.

In the midst of all this household turmoil and abuse, Bill found a place of refuge amid the rottenness of Hot Springs: a church.

By the mid–1950s, the only religious education that Bill Clinton had received came from two years in a Catholic grade school, a subject on which virtually nothing has been written. But starting in 1955, the nine-year-old began to wear a suit on Sunday mornings and walk alone to Park Place Baptist Church, about a half mile from his house. The pastor, Reverend Dexter Blevins, said that the boy was there "every time the door opened." The boy sensed, maybe from observing the behavior of the two immature adults in his small universe, that it was important that he go to church in order "to try to be a good person."[6]

His mother apparently agreed with her young son's decision to find God, encouraging him to go every week, even though she and her husband made it only on Christmas and Easter. Bill Clinton later recalled:

> I loved getting dressed up and walking down there. . . . In 1955, I had absorbed enough of my church's teachings to know that I was a sinner and to want Jesus to save me. So I came down the aisle at the end of Sunday service, professed my faith in Christ, and asked to be baptized. The Reverend Fitzgerald came to the house to talk to mother and me. Baptists require an informed profession of faith for baptism; they want people to know what they are doing, as opposed to the Methodists' infant-sprinkling ritual that took Hillary and her brothers out of hell's way.[7]

The effect of all of this was profound. The claim will shock his detractors today, but there were some who thought that the young Clinton would grow up to be a minister. His nanny, whom he called simply "Mrs. Walters," told the bright-eyed boy that someday he might make a good preaching man, like Billy Graham.[8] Though some may find the comparison amusing, Graham actually played a pivotal role in Clinton's spiritual formation. In 1958, one of young Bill's Sunday school teachers offered to take a few of the boys to Little Rock to attend a Billy Graham crusade at the football stadium at the University of Arkansas. At that time, racial tensions were high. Authorities had just closed Little Rock's schools in a last-ditch effort to prevent integration. Segregationists from the White Citizens Council recommended that Reverend Graham restrict admission to whites only. The preacher said no, stating that Jesus loved all sinners, and all people of all ethnicities needed to hear the word of God.[9]

As Clinton remembers, Graham was "the living embodiment of Southern Baptist authority, the largest religious figure in the South, perhaps in the nation." What he said carried enormous weight, and the segregationists backed down. "I wanted to hear him preach even more after he took the stand he did," said Clinton, "and the Reverend Graham delivered a powerful message." Added Clinton: "When he gave the invitation for people to come down onto the football field to become Christians or to rededicate their lives to Christ, hundreds of blacks and whites came down to the stadium aisles together, stood together, and prayed together. It was a powerful counterpoint to the racist politics sweeping across the South. I loved Billy Graham for doing that. For months after that I regularly sent part of my small allowance to support his ministry."[10]

Despite the inner strength that Bill gained from his Baptist faith, it was not until 1960, when Bill was fourteen, that he was finally able to stand up to the adult terrorizing the house in drunken rages. As an unusually large high school freshman—six feet tall, more than two hundred pounds, and bigger than his stepfather—the teen one night decided to no longer compartmentalize. He burst into his parents' bedroom, breaking down the door. "Daddy, I've got something to say to you," he warned the miserable bully. "I don't want you to lay a hand on my mother in anger ever, ever again or you'll have to deal with me." This time, Virginia finally stopped the violence—once the boys could at last protect themselves. She called the cops, and Roger spent a night in jail.[11]

While Bill's intervention put an end to the abuses at the time, soon after, in 1961, Virginia divorced Roger, only to remarry him a few months later. He succumbed to liver cancer in 1967. After Roger's death, Virginia said "I do" twice more to two more men—in all, five marriages to four different men, two of them to Roger. She was eventually Virginia Divine Cassidy Blythe Clinton Kelley.

Through this chaotic upbringing and the difficult years in a dysfunctional home, the Southern Baptist faith carried Bill. It was such a rock—mainly the only rock of reliability in his life—that Clinton at his high school graduation ceremony did something that would today shock his liberal supporters and conservative detractors: He prayed, providing no less than the benediction address. "I know that some nonreligious people may find this offensive or naïve," Clinton said when discussing the prayer later on in life, "but I'm glad I was so idealistic back then, and I still believe every word I prayed."[12]

The Partnership Begins

Yet even with the passionate Baptist beliefs of his formative years, Bill's faith would take a hiatus once he went to college in Washington, D.C., attending the Catholic institution Georgetown University.[13] Whether his decision to limit his open role in the Baptist church stemmed from a shift in his beliefs or the fact that the Jesuit institution he was attending had a much different orientation is not clear. Four years later, in 1968, Bill Clinton graduated from the prestigious college and then headed to Oxford University from 1968 to 1970 on a Rhodes scholarship. After Oxford, Bill went to New Haven, where he entered Yale Law School, and it was there that he met the bespectacled, bookish girl from Park Ridge. While their relationship began as a flirtation between two law students, both seemed to sense early on that they could benefit from each other and that together they shared a mutual ambition. It seemed as though they could survive whatever disagreements, inconveniences, troubles, and even disgraces and major transgressions, for a larger purpose of jointly transforming the world. The two influenced each other intellectually, with their experiences and their faiths rooted in such disparate pasts; together they were idealistic, but they were determined.

This is not to say that they were never in love. There is no question in the minds of those knew or observed them that Bill Clinton and Hillary Rodham were serious about each other and quite in love. They dared to break unwritten but respected Yale Law School rules prohibiting interstudent relationships, and then also shattered the mores of Hugh Rodham and Park Ridge when they decided to live together during Hillary's third year at Yale, moving into a Victorian-style house off campus.[14]

Bill seemed sure that Hillary was the girl, his life partner. Virginia was said to have been underwhelmed by Hillary, recognizing right away that she was "one of them smart ones," unlike the fun, easy girls that Bill usually nabbed for play. She was startled and not necessarily joyful when Bill told her during a visit back home during break, "Mother, I want you to pray for me that it's Hillary because if it isn't Hillary, it's nobody."[15]

During their time at Yale, the religious commitment of Bill and Hillary slowed considerably. More so than any point in Hillary's life, her years at Yale Law provided a challenge to her open practice of her faith. She did not attend church on a regular basis and was not as overtly vigilant about involving faith in her life. Michael Medved, the popular film critic and talk show host, knew Hillary well during her first year at Yale Law. Medved, who is Jewish and today is an outspoken conservative, was much more liberal in 1969 when he knew Hillary. He recalls that Hillary then seemed "just as secular as the rest of us," though he is quick to underscore that he is no authority on the subject. "I don't recall her going to church, but that doesn't mean she never did," he says. "I mean she wasn't in any way ostentatiously or visibly religious, at least during that first year."[16]

In retrospect, there seemed to be little in her life in this period that overtly tied her to church and to God. Whereas four years earlier she had entered Wellesley and still continued to attend services, this trend appears to have lapsed while at Yale.

Indeed the very language of her faith, and the way that people associated her with God, seems to have changed in New Haven. Though this had been occurring for Bill since he left Hot Springs for Georgetown, it was something new for Hillary, who was always marked by not only her religious attendance but also her open profession of faith to those around her. The record seems to back that up, with few people (if any) describing her as someone always caught in church—a clear contrast to a decade earlier, when much of her life seemed defined by her relationship with God.

While the two continued to believe and perhaps to worship in their own way, there is little evidence or discussion about how their classroom learning at this time impacted their evolving spiritual beliefs. As politics began to pull into focus more and more for both of them, religion, it seemed, would be placed on the back burner—at least, that is, until it became necessary.

Radicals

It was no secret among those who knew Hillary that over the course of her time at Wellesley she had moved steadily to the left. Upon her arrival at Yale, this was a trend that continued, as she was further influenced by a rapidly growing antiwar movement and the radicalization of the American left that was sweeping the country.

During her time at Yale, Hillary met Marian Wright Edelman, a

Yale Law grad who was the first black woman admitted to the Mississippi bar. From 1963 to 1967, Edelman directed the NAACP Legal Defense Fund. The next year she started the Washington Research Project, which soon became the Children's Defense Fund (CDF). In the summer of 1970, Hillary interned with Edelman at the CDF, an experience that proved to be a perfect fit for her. With its broad goals and expansive ideals, the CDF was the ideal engine for remaking society. The cause of "children's defense" was an all-encompassing one that overlapped with civil rights, social justice, and the class struggle, all under the headline of protecting the welfare of children. That net became so wide that CDF would find itself involved in everything from minority kids to Greenpeace to the nuclear-freeze movement.

The CDF proved to be a solid foundation for Hillary, who came to see how the organization used this myriad of causes to push a Democratic agenda of federal welfare programs, income tax increases, and the redistribution of wealth. But while Edelman and the CDF were solidly on the left, they were not at the extreme end of the spectrum. By 1971, however, such fellow travelers were not outside Hillary's orbit.

Throughout the summer of 1971, Hillary spent her time in Oakland, California, where she was hired as an intern in the law offices of Robert Treuhaft, husband of the British-born Jessica "Decca" Mitford, the onetime muckraking journalist; they lived near Saul Alinsky. In San Francisco, as the *New York Times* put it, "Mr. Treuhaft started a radical law firm that specialized in fighting every kind of discrimination and social injustice."[17] Treuhaft and his wife were well known throughout the Bay area and around the country. Both had been members of the Communist Party USA, during which time they were frequently denied passports and investigated by government officials. (Hillary's introduction to the couple probably unfolded through a law professor with whom she was close—Yale constitutional scholar Thomas I. Emerson, known as "Tommie the Commie," and another reported CPUSA member.) Eventually, they left the party in 1958, two years after absorbing the "staggering blow" of Nikita Khrushchev's revelations of Stalin's massacres.[18]

Although Hillary's new boss had abandoned the Communist Party, he did not forsake the party's goals of social revolution. By 1971, Treuhaft welcomed Hillary Rodham with open arms to his latest causes, which at that point included the free speech movement, the Black Panther Party, and the draft-resisting Oakland Seven during the Vietnam War.[19]

While it's unclear to what extent Treuhaft continued to uphold Communist Party doctrine and view religion as the "opiate of the masses" when Hillary was working for him, the mere fact that Hillary found herself at his law office during that summer speaks volumes about the distance that her faith and her politics had traveled. Though Treuhaft pursued an aggressive social agenda that fit in well with Hillary's underlying Methodist ideals of economic social justice, his far-left ideals bordered on a political line more extreme than anything with which Hillary had previously associated. Whatever his religious beliefs (or lack thereof), she seemed willing to put hers on hold, or at the very least bite her tongue for a summer, in the name of the various causes on which she had begun to focus.

This became an increasingly common trend for Hillary during these years when she was finding herself politically. As the Vietnam protests escalated and the country grew more and more divided, the decline of her open spirituality that had begun at Yale seemed to continue. It was not so much that her Methodist goal of social change was different; on the contrary, her youthful ideas of helping society were now being pushed to their extreme limits. In the process, she was drifting from the religion that had helped her to arrive at those goals in the first place. This tenuous balancing act between faith and politics would tip further in favor of politics as the 1970s progressed, and more and more Hillary found herself overlooking ideas about faith in favor of ideas about politics.

This included her view on Jews and the nation of Israel. In July

1973, Bill, who was already sizing up his political prospects, called upon a politically well connected friend in Arkansas, and Hillary accompanied him. According to author Christopher Andersen, as their car approached the house, Bill and Hillary noticed that a Jewish mezuzah was affixed to the front door. The owner made clear that he was proud of the symbol, but to the astonishment of the gracious host, when Hillary saw the mezuzah, she reportedly refused to get out of the car.

In the words of the friend and another eyewitness interviewed by Andersen, Bill said, "I'm sorry, but Hillary's really tight with the people in the PLO in New York. They're friends of hers and she just doesn't feel right about the [mezuzah]."[20]

"Do you mean to tell me," Bill's friend shot back, "that she is going to be part of Yasser Arafat and all those people?"

Bill, embarrassed, shrugged and said, "Hillary really backs the PLO and doesn't like what Israel is up to."[21]

If accurate, rather than suggesting nascent anti-Semitism, this alleged incident more likely reflected the primacy that politics had achieved in Hillary's life. Though her faith told her that she should love all people, regardless of color or creed, this principle was colliding with her politics.

Big Ideas

In 1973, Hillary graduated from Yale Law School and went to work as an attorney for Children's Defense Fund in Cambridge, where she would stay until joining the staff of the House committee to impeach Richard Nixon in January 1974.

It was there that Hillary's radicalism was manifested. Most of those looking to impeach President Nixon fixed their gaze on Watergate and the cover-up. Hillary, however, had a bigger view. According to

Joyce Milton, she tried to add Nixon's bombing of Communist supply lines in Cambodia—a crucial conduit for the Vietcong—to the list of articles of impeachment. Nixon's resignation spared Hillary the cup of victory, though she celebrated his resigning in disgrace.

While Nixon's lifetime of elected office was ending, Bill Clinton's was just beginning. Fitting for a man who seemed born a politician, it took Bill only months after likewise graduating Yale Law in 1973 to launch his first campaign for national office, running for Congress in Arkansas's Third District, where he got help from Hillary and both of her brothers but was eventually defeated. Despite this setback Clinton was not discouraged, and instead of adopting a defeatist attitude, he stepped back from the experience, wisely reassessing and lowering his expectations.

Because of Bill's 1974 bid for Congress, he and Hillary had uprooted from New Haven for Arkansas. After the move, both he and Hillary joined the law faculty of the University of Arkansas at Fayetteville. Though Hillary felt like a fish out of water in Arkansas, the small state was ripe for the two to make a big splash. Hillary had become tenacious about her politics, and whereas Bill was more politician than ideologue, Hillary was more the type to jump off a cliff with her flag flying in the wind.

During this period, Hillary reached what would become a monumental crossroads in the relationship between her religious and political views. On January 22, 1973, a group of men on the U.S. Supreme Court legalized abortion in a case called *Roe v. Wade*, with the majority opinion written by Justice Henry Blackman, a fellow Methodist. At some point in this period, though precisely when is a subject of great speculation, Hillary became not only a supporter of this decision but one of the country's most vehement advocates for abortion rights. How, exactly, this happened is difficult to pinpoint, though one important source has insight.

In 1974, Hillary met William F. Harrison, a prominent abortion doctor in Arkansas, who became her gynecologist and friend. In a series of interviews for this book, Harrison shed some light on the development of Hillary's pro-choice stance and how she made up her mind to support the controversial Supreme Court decision. Harrison first met Hillary and Bill in Fayetteville, after she and Bill had joined the University of Arkansas legal faculty. The first time he encountered Hillary it was as her physician; a short time later, however, he met Bill at a local restaurant, and Harrison's friendship with the couple was sealed.

Over time, he came to know them personally as well as through his practice, though Harrison is quick to point out that Hillary never saw him for an abortion. The specific reasons for their introduction have been a source of speculation around Arkansas ever since that first meeting, but Harrison says that he met Hillary simply as a result of her yearly ob-gyn exam—presumably, her first checkup since moving to Arkansas. "Hillary never saw me for an abortion. I don't know of any abortion that Hillary ever had. And I would be shocked if she had."

This is an important point in trying to discern Hillary's dedication to the abortion issue, since it would mean that Hillary's unflagging support does not stem from a personal experience in which she had the procedure—a point of immense speculation among Hillary's political observers and critics.[22] Rather, Harrison estimates that a central reason for Hillary's pro-choice stance is that she is a product of an age "where she would have had friends who had illegal abortions. . . . I am sure that was part of it. And, you know, her church was supportive of that opinion."

Harrison says that when he met Hillary in 1974, she was already steadfast in her support of *Roe v. Wade*. In fact, Harrison sees Hillary's upbringing as a devout Methodist as no reason to believe she would be against abortion. "Hillary [is] a Methodist and I was raised a Methodist. The Methodist church [is] very strongly pro-choice."

Permanent Partners

As Hillary continued to develop these lifelong positions, she also sought to create a lifelong bond with the man in her life. After almost three years of living together, Hillary and Bill set a wedding date for October 11, 1975. However, now as they prepared to take this major step together, the couple first had to deal with their religious differences.

Though both were of course Protestants, there was a big difference between a Midwest Methodist, who, by now, was really a Northeast Methodist, and a Southern Baptist. Despite the fact that in the last few years their devotion had been steadily ebbing, it did not make the matter of reconciling their faiths to each other any less difficult. "We, of course, think the most important thing is your personal relationship with God," said Hillary later, "and the denomination you belong to is a means of expressing that and being part of a fellowship."[23]

Even with this open-minded view toward each other's faith, it seemed curious that the couple did not marry in a church—even a neutral or nondenominational church.[24] Instead, they were married in the living room of a house they purchased at 930 California Drive in Fayetteville, Arkansas, in a private ceremony that included academic and political friends and family. All the Rodhams were present, as were Bill's mother and brother, who served as best man. Hillary wore a Victorian lace dress, and one can only wonder what Hugh Rodham, the stern, traditional Methodist, felt as he stood there in the living room and watched his sole daughter start a new life.[25]

The minister who presided over the ceremony was a local Methodist pastor named Vic Nixon, whose last name compelled the political Hillary to quip, "Never thought I'd be married by a Nixon!"[26] Hillary insisted that the minister be a Methodist, and so Bill had contacted Nixon two months earlier in preparation for the wedding to ask if the minister would be willing to marry the two strangers.

Nixon said he would be happy to do so, but that he would like to sit down and visit with them first.

Bill had some concerns on that score, perhaps fearing whether the fact that the two had lived together would be an issue. "[A]re you going to ask us questions?" Bill said guardedly. Nixon replied, "Well, I don't know. I may ask some." Nixon suspected: "I think Hillary wanted to know what the questions were so she could study for the visit."[27] In the end, the visit was brief and pleasant; they met and talked, though Nixon has always kept quiet about the content of the premarital discussion. Aside from choosing the faith of the minister, Hillary insisted on another liberty: She wanted to retain her maiden name. Bill acquiesced, but in the years ahead this choice, along with the decision not to marry in a church, would lead some of the more religious and traditional folks in Arkansas to question if the two were actually married.

During their first years as husband and wife, politics was paramount—with no children and scarce churchgoing. Yet, while Hillary devoted most of her lawyerly energies to left-leaning efforts and women's causes, such as the Equal Rights Amendment, her interests varied, and on at least one occasion she involved herself in a legal case that would both surprise and please many of the Christian faith. While living in Fayetteville, Hillary got a call from a female jailer regarding a woman who had been arrested for disturbing the peace by preaching the Gospel on the streets of nearby Bentonville. The judge wanted to send the woman, a California native, to a state mental hospital, and the jailer asked Hillary to come to the prison right away, believing that the lady was not crazy but simply "possessed by the Lord's spirit."[28]

When Hillary got to the Benton County prison, she encountered what she described as a "gentle-looking soul" wearing an ankle-length dress and tightly clutching a well-worn Bible. The woman explained that Jesus Christ had sent her to Bentonville to preach, and once she was released, she would continue her mission. When Hillary learned the woman was from California, she persuaded the judge not only to let the lady go but to purchase her a bus ticket home rather than committing her to an asylum. She then convinced the woman that "California needed her more than Arkansas."

Despite her transgressions, Hillary, it seemed, had not strayed so far from her roots that she would ignore a chance to help a sister in Christ who faced difficult and unjust circumstances. Though she had stopped attending church, there were signs that Arkansas would have the potential to reinvigorate her faith, as the very culture of the state and of the South afforded a greater opportunity and openness for worship than the other places she had lived since leaving Park Ridge.

Still, she placed a distance between the outside world of the state and her own set of interests. Hillary continued to pursue her Democratic causes, even though the state as whole was trending closer and closer to the right. As the 1970s progressed, this would factor into her young husband's political career with increasing regularity. Nevertheless, the question remained: How far was she willing to bend her own values and beliefs in the name of her husband's political career?

CHAPTER 5

The First Lady of Arkansas

The year 1976 was decisive in foreshadowing the careers that Hillary and Bill would have for the remainder of that decade and the entirety of the next. For Bill, 1976 marked the first time he was elected to public office when the Arkansas voters selected him attorney general of the state. For Hillary, that year was the start of an entirely new turn in her practice of law, when she began fifteen years of work as an attorney for the Rose Law Firm in Little Rock.

Politically, 1976 was small potatoes for Bill, who recognized the need to hold the attorney general post before he could take the next step to the governorship. Bill did not waste time before making an attempt, and in 1978 he ran against Republican challenger A. Lynn Lowe for the governor's office. The election was a telling one. The thirty-two-year-old Clinton trounced Lowe in a landslide, 63 percent to 37 percent.

The young politician and his wife had just won their second election, and it would be a historically significant one. While the two had been training for the spotlight for years, now that they achieved it they found themselves in unfamiliar territory. There was a multitude of questions about how they would react to the new life in front of them—and for Hillary in particular, there were questions about how her Democratic ideals, which were becoming increasingly secular, would fly in the religious state of Arkansas.

Social Justice with the Law

As first lady of Arkansas, Hillary retained a project close to her heart: the CDF. After leaving CDF, she had remained in touch with Marian Wright Edelman. Now, with her husband becoming governor, she was invited to serve on the board of directors of CDF, which she did eagerly, continuing this association throughout the 1980s until 1986, when she was promoted to chair of CDF. (Ultimately, she left in 1992 to pursue her husband's presidential campaign.)

Yet the group that really showed Hillary's political colors—and nearly got her into later political trouble—was the Legal Services Corporation (LSC). Hillary was appointed to the board of LSC in 1978 by President Jimmy Carter, and would remain there until 1982, at one point becoming chair of the board. From that perch, she set out to change the world by means other than elected office. In her case, the revolution would be won by the LSC through the courts. What was more, the LSC offered a convenient opportunity, since the cause could be underwritten by American taxpayers.

The LSC had origins in Lyndon Johnson's Great Society, though it was not formally created until 1974 when Richard Nixon signed it into law. The justification lay in the constitutional right of every American to have legal representation, even when unable to afford a lawyer. Of course, historically, attorneys were provided by local bar associations, legal societies, or the local government; the poor got representation through local public defenders. For Hillary and other like-minded people, however, this was not satisfactory; to them the only way to achieve effective representation was with the involve-ment of the federal government. This is where the LSC stepped in—to provide attorneys from the federal system to represent the needy in a variety of cases ranging from the criminal to the civil.

Thus, the LSC became a magnet for liberals fresh out of law school and committed to changing society through the court system. What the children of the 1960s could not achieve through the ballot box, they hoped to now gain by the mighty arm of the lawsuit. Moreover, the potential to multiply the results appeared limitless. Individual criminal cases filed by plaintiffs could be taken well beyond one little courtroom in one small town; they could be pushed up through as many levels of courts as possible, and maybe all the way to the U.S. Supreme Court.

Once created, the task was to staff the LSC with similar-thinking people. The Carter administration did just that, with Hillary a key addition. By the time Hillary was appointed to the board of directors in 1978, the LSC had already ballooned into a $200-plus-million federal program to mete out social justice.

With Hillary as a member of the premier law firm in the state of Arkansas, the Rose Law Firm, it just so happened that these LSC activ-ities were perfectly poised to expand legal services across the nation in a way that had the potential to generate a large amount of business for Rose—especially in regional cases where business and individu-als would turn to them for guidance. Peter and Timothy Flaherty, who have done the best research on this subject, note that Hillary's appointment to the LSC created "a little-noticed controversy" when it was submitted to the U.S. Senate for confirmation. The Arkan-sas first lady's successful efforts to expand legal services in Arkansas meant an increase in litigation against individuals, businesses, and government bodies in the state. Hillary was not only poised to build legal services, add the Flahertys, she now had the potential to serve as a "rainmaker" for the Rose firm.[1]

This was a conflict of interest of which Hillary was obviously aware but refused to openly acknowledge, fueled by her strong belief in the moral center of the cause. Undeterred, she and her associates did anything but fly under the radar. They pursued very bold lawsuits that placed them squarely in the spotlight of public attention and outcry. The Flahertys point to suits like suing the school board of Ann Arbor, Michigan, in an attempt to force teachers to teach "Black English" and filing suits in several states trying to force the federal government (through Medicare) to pay for sex-change operations.[2]

More explosive, as the Flahertys note, was the effort by LSC forces in Illinois to undermine the Hyde Amendment, the legislation by Congressman Henry Hyde (R-Ill.) to ban federal funding of abortion. Many Democrats were sympathetic to this law; President Jimmy Carter, himself a pro-life Democrat, even supported the amendment. Carter had, however, turned the LSC over to Democrats whose politics on this issue were completely contrary to his own. Committed to maintaining what they believed was a basic right for all women— including those who could not afford it—they mounted a legal challenge to the amendment.

In 1980, a major threat to the LSC emerged: Ronald Reagan. Reagan had been elected president in November 1980, and the LSC crusaders feared the game was up. By the estimation of the hard left, Reagan and his fellow Republicans were ready to sideline their vehicle for social change. In the end, it was only through a remarkable campaign of self-preservation that Hillary and her allies were able to salvage the LSC from Ronald Reagan's domestic ash heap of history. In doing so, however, the subsequent actions of certain staff were legally suspect and ethically outrageous, culminating in Senate hearings and GAO investigations into the organization. Ultimately Reagan replaced the Carter board with his own appointees in 1982, and somehow Hillary escaped unscathed.

During this time, the role of Hillary's faith was not unclear: The social justice that she was seeking through the LSC was consistent with the religious-left progressivism she had begun to learn twenty years earlier under Don Jones. And, all along, her moral absolutism was evident in her actions; she possessed a kind of religious sureness and constancy that lent inflexibility to her cause. In short, the LSC was for Hillary a vehicle to gain that old-time social justice through legal justice.

Reelection Woes

Because of the brief two-year term of the Arkansas governorship, Bill was up for election again in 1980, and this time the election proved much more challenging.

Much of Clinton's difficulty with the campaign stemmed from rioting that took place at Fort Chaffee, Arkansas, during the summer of 1980. This was the period of the infamous Mariel boatlift, in which President Jimmy Carter welcomed any and all Cuban émigrés that Fidel Castro desired to export from his island prison. Castro took the opportunity to cleanse his country politically and socially, opening up the doors to many of his jails and asylums and shipping their contents off to the United States.

In the weeks that followed, more than one hundred thousand "Marielitos" washed upon America's shores, and Jimmy Carter, who was unprepared for the dramatic influx, did not know where he was going to temporarily place all of them. In the fray, he telephoned Governor Clinton, who said he would be willing to detain some of the escapees at Fort Chaffee, where they could be assimilated.

The detainees, some of them sporting tattoos denoting the number of people they had killed in Havana, overflowed the grounds. The Arkansas Ku Klux Klan drove to Fort Chaffee to express its disapproval of the state's newest group of tourists, and as a result, a riot ensued between shouting Cubans inside and KKK belligerents on the outside, with various projectiles tossed back and forth. Soon, law enforcement arrived to try to hose down and disperse the crowd, but for Clinton's reelection, it was too late. The entire fiasco was captured by state news organizations for everyone to see, and Arkansans, led by Republican gubernatorial candidate Frank White, naturally asked why such a debacle had been permitted to happen in their state.

A few months later, Bill Clinton was defeated, losing to Republican Frank White in a close race, 52 percent to 48 percent. But he was far from finished; he lived for politics. And he and his wife began strategizing about how to return him to power.

The plan they came up with called for the first of innumerable Hillary makeovers in the years ahead, all with the intent of toning down her more liberal credentials and traits. During her two years as the first lady of Arkansas, she had not been embraced in Arkansas, with many constituents viewing her as a hippie from the Northeast, a "women's libber" who was not the kind of first lady to which they were accustomed. She had kept her maiden name, thereby—in the eyes of many Arkansans—calling into question the governor's manhood.

In early 1981, as Bill geared up for a campaign to return to the governor's mansion, Hillary Rodham did an about-face: She became Mrs. Clinton. Though this appeared to be a shocking turn of events, some biographers of Hillary say that her new name appeared only on her business stationery. She remained Hillary Rodham when she voted, paid taxes, or filed legal papers.[3]

During these years, between gubernatorial terms, the Clintons returned to church, albeit separately. Biographer David Maraniss notes how the demands of "their partnership" led religion to play "an increasingly important role" in the lives of the Clintons through the 1980s. Both found that their faiths "eased the burden" of their high-profile lives, "sometimes offering solace and escape from the contentious world of politics, at other times providing theological support for their political choices." As success continued to follow them, their religion continued to follow right alongside.[4]

Skeptics saw a correlation between the Clintons seeking a church and seeking votes in this Southern state, and few biographers have failed to raise an eyebrow about this apparently abrupt about-face in the couple's attention to religion. Whereas the two seemed to have sworn off regular church attendance for years, suddenly they were back in the pews. For many, the connection between this change in behavior and Bill's campaign loss in 1980 was too convenient to dismiss as coincidence, but in reality, the Clintons were equally likely following the common trend of young people who leave the faith temporarily in college and then return as parenting adults moving toward middle age.[5] Such a return to religion could be expected for Hillary and Bill, since both had been devout as kids and teens. Moreover, on February 27, 1980, Hillary had given birth to the couple's only child, Chelsea, named for a Joni Mitchell song, "Chelsea Morning."

But while they agreed to return to church, a split remained over denomination. Hillary was not about to abandon Hugh Rodham's Methodist roots, and Bill remained committed to the Baptist faith that had saved him from the home of Virginia and Roger Clinton.

In Little Rock, Hillary opted for First United Methodist, a wealthy congregation filled with young professionals, many of them upper class, with a special attraction to attorneys—seventy-six of its members were lawyers, including many of the top lawyers in Little Rock (the local bar association held its monthly meetings there).[6] Located on 723 Center Street, First United Methodist was Hillary's kind of church, considering its rich history of involvement with civil rights and social justice in the Little Rock community. After it was founded in 1831, whites and blacks attended services together during the church's first forty-five years, except for two years during the Civil War, when the building was requisitioned as a hospital by the Federal Army. With this pedigree of progressive thought and socially conscious action, it was clear that this was going to be the right place for Hillary to worship.

It was no doubt Hillary's intention to do her part to further the parish's long history of social dedication. While she was a member of the church during the 1980s, the congregation purchased a large building at Eighth and Spring streets to expand its day care center into what became the Gertrude Remmel Butler Child Development Center. Nevertheless, Hillary's exact role in this endeavor was difficult to ascertain, as nearly every single person who worshipped with her in Little Rock and was contacted for this book refused to be interviewed.

Regardless of her specific involvement in the center's creation, she clearly supported the center, which was consistent with her deep interest in child care. She personally donated funds to the center, which today serves more than three hundred children in full-time and after-school child care, children of working parents in downtown Little Rock.[7] Known as the "CDC," the Child Development Center is the largest state-licensed child care facility in the state of Arkansas, housing a huge staff of more than seventy.[8] The facility in many ways reflects the vision for communal support that Hillary would later outline in her book *It Takes a Village*.

Beyond the building of the CDC, Hillary was very involved in the church and took a leadership role. She served on the administrative board and performed pro bono legal work, acting as legal counsel for the local bishop, Richard B. Wilke, who described her as a "vibrant and vital part of the life of [the] congregation."[9] This legal work extended outside the Little Rock church: For a time she was the lawyer for the Methodist Conference of Arkansas.

The First United Methodist Church of Little Rock was then and remains today a liberal congregation, certainly relative to the rest of the South, and reflective of the direction of the denomination nationally. More recently, in 1998, it welcomed its first female pastor, the Reverend Jeanie Burton, who was appointed by Arkansas's first female bishop. This change would have thrilled Hillary, as would others relating to women: At the time of the writing of this book, two newsletters were posted on the church's Web site, the first of which gave a pitch for a women's conference in California, which featured political and social "workshops" on women and children; the second and most recent newsletter promoted a guest speaker for the "Unity Sunday School Class" at the church—a mother who was coming to the church to discuss how "religious beliefs undermined . . . and created significant tensions" between her and her lesbian daughter and "ended in a tragic loss." The newsletter also plugged an upcoming program at Hendrix College, a Methodist institution in nearby Conway, Arkansas, where the topics of study were globalization, the "peace" of the three "Abrahamic" religions, and the "richness and diversity" of the cultures of India and Pakistan. In an interesting irony, the current newsletter also welcomed to the congregation a young woman who would be working with the church youth group, a minister-to-be, headed off to graduate school at Drew University—the seminary of a man named Don Jones.

A Church for Bill

While Hillary found her spiritual home in Little Rock with relative ease, picking a church for Bill was far more controversial because of political suspicions that arose about his motivation to start attending services. In the end, Bill opted for Immanuel Baptist Church of Little Rock, an interesting choice to be sure, and one that rightly raised questions. As everyone in Arkansas knew, Sunday services at the church were televised and had been since the early 1970s, meaning that Governor Clinton could now be seen at church by Arkansas voters every Sunday morning.[10]

Or could he? This presented another complication since the logical place for Clinton would be among the congregation, not behind the pulpit, but here again, Clinton defied expectations, opting instead to join the choir, where he occupied a seat that was perfectly positioned directly behind the pulpit in full view of the TV camera. To his supporters, it was reassuring to know that their governor was a pious man who was worshipping alongside them every week, though even those who liked Clinton could not help but raise an eyebrow. To his critics it was a shrewd and careful orchestration, a public display designed to reveal him as a churchgoer before the hundreds of thousands of viewers (and voters) who tuned in every week.

Some politicians—even would-be president Ronald Reagan—stopped attending church once they reached higher office, in part because they did not like to be stared at while they worshipped. For Bill Clinton, this may have been precisely the objective.[11] After all, it was not as if Immanuel was the only Baptist church in town: There are more than sixty Baptist churches in Little Rock.[12] The one that Clinton picked was not exactly chosen for convenience, as it was an eight-mile drive from the governor's mansion; Bill Clinton presumably passed dozens of other Baptist churches along the way. Also, the pastor was not exactly the Clintons' cup of tea: Worley Oscar "W. O." Vaught was a conservative, a fact that Bill had known ahead of time, as had all of Arkansas. This, however, was no deterrent to Bill selecting Immanuel Baptist. Despite the myriad reasons for him not to join the congregation, ultimately none of these factors could dissuade him from selecting Immanuel Baptist, and in 1980 Clinton joined the church—and the choir.[13]

Clinton himself quipped that many Arkansans had a hard time imagining what a liberal like him was doing in Vaught's congregation—"I was this young firebrand and he was an old conservative minister."[14]

Even in the face of much speculation, Bill has avoided remarking on whether there was a political motivation for picking Immanuel. In Clinton's defense, the church was a remarkable place of worship, overlooking the capitol and housing its four-thousand-member congregation—the largest in the state. In many ways, it should not have been a surprise that Bill was drawn to Immanuel, even though the state's citizens watched the services on their TV sets at 11 A.M. each

Sunday. As more than one biographer has noted, it was during those services that Clinton found faith and fortitude in dealing with the rigors of his job. Like other Baptists, each week he carried his worn and well-read Bible to church, opened it up inside, listened closely to sermons, and sang loudly.

Clinton said he "admired" Vaught and his "careful" teaching of the Bible. "He believed that the Bible was the inerrant word of God but that few people understood its true meaning," said Clinton. "He immersed himself in the earliest available versions of the scriptures, and would give a series of sermons on one book of the Bible or an important scriptural subject before going on to something else." Clinton said that he looked forward to his Sundays in the choir, "looking at the back of Dr. Vaught's bald head and following along in my Bible, as he taught us through the Old and New Testaments."[15]

While there was no doubt that his time at Immanuel was helpful for him spiritually, much of the evidence seems to suggest that Clinton's choice of Immanuel Baptist was also the first of many times that he would use church appearances for political opportunism. In the end, his time at Immanuel helped break down a crucial political as well as personal barrier for him; it would only be a matter of time before he would begin campaigning in churches, using public displays of his faith to score him additional political points.

On the other hand, biographer Nigel Hamilton says that it was Hillary, not Bill, who suggested that he attend Immanuel for political reasons. Hamilton writes that Hillary remained doggedly certain that Bill could eventually become president if he was reelected governor, and so together they both agreed that Immanuel was the right choice.[16]

Clinton's own words at least partially substantiate Hamilton's conclusion, as he himself points to Hillary as the reason he chose First Immanuel Baptist Church and joined the choir that seemed so conveniently within camera range—though Bill says that Hillary prodded him for spiritual rather than political reasons. "[M]y wife persuaded me to start going there and to join the choir," said Clinton categorically, adding: "She said I obviously felt the need."[17] More than once, including in his memoirs, he said he joined Immanuel and the choir "at Hillary's urging," always adding: "Hillary knew that I missed going to church."[18]

While Clinton has always freely acknowledged that he had quit going to church on a regular basis at the start of college, he maintained that the second phase of his faith began shortly after he was elected governor for the first time. "In 1978," he said later, "when I got elected governor, it was important to me to have a dedicatory service."[19] Yet despite this remark, it was still another two years before Clinton would start attending regular services and begin practicing his faith once again.

The Influence of W. O. Vaught

Whatever the motivations for Bill's choice—only God and the Clintons know the true answer—more important is what Clinton learned from his conservative minister at the church. David Maraniss writes of the profound influence that Dr. W. O. Vaught had on Clinton and many of his eventual policy stances. Beyond the preacher's sermons, Bill shared frequent conversations with Vaught, who took time to guide the young governor spiritually and help him balance the reemergence of his faith with his political life.

In addition, Clinton was also spiritually guided by different members of Vaught's flock, some of whom worked with the governor. Each morning when Bill reached his office, there was a Bible quote on his desk, placed by a fellow attendee of Immanuel Baptist—his personal secretary, Lynda Dixon.[20] It was the custom of Clinton and Dixon to begin the governor's workday by reading and discussing the passage. Here were two of Vaught's students continuing the Sunday practice, meaning that Bill Clinton was not strictly a "Sunday morning Christian."

Yet Vaught's influence was much more direct, and in a way that would no doubt intimidate the more secular-minded elements within the Democratic Party today. According to Betsey Wright, one of the governor's closest aides, Clinton each day included his pastor among his regular round of phone calls, tapping him as a resource in considering policy decisions. Clinton found himself calling upon Vaught constantly during this crucial period when he worked through issues that would extend from his governorship to an eventual American presidency. Betsey Wright remembers the governor often remarking after a Vaught phone call, "Well, Dr. Vaught told me such-and-such."[21]

According to Maraniss, this close relationship between preacher and his congregant became very influential for Clinton on two issues in particular, in which Vaught had a significant impact on how Bill Clinton perceived the religious implications of his decisions.

The first was on the death penalty, on which Vaught took the initiative to reach out to Clinton when it became apparent that the governor would soon be setting dates for executions and deciding which death-row inmates would or would not be spared. In 1976, when Clinton ran for attorney general, he told conservative Southerners that he advocated capital punishment. When he became governor, in the early 1980s, the lives of certain incarcerated citizens once again lay directly in his hands, but whereas during his first time in office he did not have spiritual guidance, now he had a pastor who could sense that Clinton was troubled. Vaught called, and Clinton quickly responded by inviting the pastor to the governor's mansion to advise him on the subject.

Clinton asked the Baptist minister if it was biblically permissible for him to execute a man, and Vaught told him that the death penalty was not prohibited in the original translation of the Ten Commandments. He went further into the subject, examining the Old and New Testaments, the Hebrew and Greek. The final decision would be Clinton's, noted Vaught, but he "must never worry about whether it's [the death penalty] forbidden by the Bible, because it isn't."[22]

The second major issue on which Vaught helped to guide Clinton was abortion. In this case, Clinton sought Vaught, and again on this question, Clinton was reportedly troubled, expressing a deep personal ambivalence. Maraniss reports that Clinton consented to the pro-choice argument intellectually, especially since he surrounded himself with devoutly pro-choice women, including Hillary. Indeed, Hillary's ob-gyn, William F. Harrison, the abortion doctor, stated that Bill was firmly pro-choice as early as 1974; in fact, Harrison was angry that Clinton did not speak out more forcefully in support of legalized abortion throughout his governorship—which the Fayetteville doctor ascribes to Bill straddling the fence to appeal to voters.[23]

Nonetheless, something inside Bill Clinton—his conscience, presumably—was prompting questions, perhaps even second thoughts about his stance on the issue. Maraniss says that the governor was struggling over the definition of human life. Could Vaught provide some insight from the Hebrew and Greek?[24]

Although many would presume to know what Vaught's response was after consulting the text, in reality Vaught's reported reaction was interesting and quite surprising. He was one of the leading abortion opponents among Little Rock clergy, but said he shared some of Clinton's ambivalence, having personally witnessed "some extremely difficult" pregnancy cases as a pastor. He was not convinced that the Bible forbade abortion in all circumstances. What he most likely meant by this was that there were no literal biblical passages condemning or describing precisely what a woman should do in each situation.

In Maraniss's account, the minister went to his Bible to revisit and reconsider, after which Vaught determined that in the original Hebrew, "personhood" stemmed from words translated as "to breathe life into." Thus, he averred, the Bible would define a person's life as beginning at birth, with the first intake of breath. He reportedly told the governor that this did not mean that abortion was right, but he felt that one could not say definitively, based on Scripture, that it was murder.[25]

Vaught's guidance proved instrumental. Says Maraniss, "In all of his discussions about abortion thereafter, Clinton relied on his minister's interpretation to bolster his pro-choice position."[26] In short, then, this conservative, pro-life Baptist minister, based on Maraniss's account of the situation, helped steer Bill Clinton into the pro-choice position from which he never again wavered.

Though Pastor Vaught may have felt in his heart that abortion was wrong, it was up to him alone—possibly in consultation with other "Bible Protestants"—to study his Bible to assemble the collection of scriptural references that he used to guide Clinton's decision. In the early 1980s, the decentralized system of Vaught's denomination meant that it was up to Vaught alone to make these determinations for his four-thousand-member congregation until his Baptist church could vouch for the biblical soundness of the pro-life position. For tens of millions of other Protestants in the United States, the issue would become the single greatest moral-social issue.

Still, despite Vaught's emphasis on his personal scholarship, his conclusion and his advice to Bill were surprising given the other spiritual red flags that abortion was raising around the world. At the time of Vaught's counsel, the Catholic Church was already strongly committed to the pro-life position. It had come to that position not through invocation of a single Bible passage that featured the word "abortion," but through a careful meeting of the minds between the Church's highest and most careful scholars. Indeed, had Vaught looked beyond the immediate confines of his personal study to what other scholars of the original Greek and Hebrew were concluding, he might have come to a far different conclusion and swayed Clinton to a pro-life position.

Ultimately the Protestants who followed Vaught would conclude that the unborn human was in fact a life, and terminating that life constituted murder or killing, in this case of a helpless innocent, which the Bible forbids. As the 1980s progressed, many Protestants began identifying Bible passages and stories about God knowing and weaving humans in the womb, about the humanity of life in the womb—about Jacob engaging in conflict in the womb, about Tamar's twins in the womb, and, most notably, about John the Baptist leaping for joy in the womb when encountering the presence of the Christ child in Mary's womb. Perhaps most striking, a passage in the Old Testament's Book of Ecclesiastes (11:5) speaks of the "breath of life" fashioning the human frame in the womb, a verse that should have been instrumental to Vaught's conclusions.[27]

The revelations, however, came too late for Clinton. There was just enough of a delay for him to end up firmly in the camp of his wife on the abortion issue, and so from this base belief that his minister reportedly helped to carve, Clinton would go on to become the most pro-choice president in history.

The Pilgrimage to Israel

During this time Vaught began having an effect not only on Bill but on Hillary as well, and in December 1981, while the Clintons campaigned to win back the governorship, Vaught approached them with an interesting proposition. Since 1938, ten years before the state of Israel came into being, Vaught had been venturing to the Holy Land, and now as Bill and Hillary found themselves struggling spiritually and politically to put Bill back in the governor's mansion, the couple decided to accompany Vaught on the trip.[28]

According to Bill, Hugh and Dorothy Rodham were big supporters of this "pilgrimage with Hillary," and the grandparents even ventured to Little Rock to stay with baby Chelsea during the trip. On some level, Hugh must have relished the prospect of helping Hillary to reconnect with her spiritual roots after so much time of separation. Indeed, Hillary biographer Norman King claims that the trip was made "largely because of Hillary's insistence." King said that the Clintons viewed the trip as both a religious pilgrimage and a chance for the governor to retool and refresh—a change of scene and boost of morale.[29]

While on their journey, the Clintons spent most of their time in Jerusalem, retracing the steps that Christ himself walked and meeting with local Christians. Bill noted that they went to the Sea of Galilee, "where Jesus walked on water." They saw the Church of the Holy Sepulcher and toured the city of Jericho. They saw the spot where Christ was crucified, and visited the small cave where Christians profess that Christ was buried and arose on the day now recognized as Easter Sunday.[30] The entourage also went to two sites sacred to other faiths: the Western Wall, holy to Jews, and the Dome of the Rock, where Muslims believe Mohammed ascended to heaven to meet Allah.[31]

Through their travels, the Clintons also contemplated Armageddon and the mighty clash of civilizations to which the land had borne witness. Bill and Hillary went atop the city of Masada. "[A]s we looked down on the valley below," said Bill, "Dr. Vaught reminded us that history's greatest armies, including those of Alexander the Great and Napoleon, had marched through it, and that the book of Revelation says that at the end of time, the valley will flow with blood."[32]

In contrast to the anti-Israel version of Hillary portrayed during parts of the 1970s, some sources claim that this trip gave Hillary an inspired appreciation for the state of Israel, and, if so, it may have mitigated her alleged pro-PLO sympathies, giving more balance in her perspective. On this, Norman King cites Sarah Ehrman, a friend of the Clintons from the anti-Nixon, McGovern days: "Bill and Hillary understood the profound effect that Israel has on American Jews and around the world and share a feeling for the security and stability of the State of Israel."[33]

Bill himself said that Reverend Vaught was instrumental in his own attitude toward Israel. "I have believed in supporting Israel as long as I have known anything about the issue," Clinton said later. "It may have something to do with my religious upbringing. For the last several years until he died, I was very much under the influence of my pastor." He also says of Reverend Vaught: "He was a close friend of Israel and began visiting even before the State of Israel was created. And when he was on his deathbed, he said to me that he hoped someday I would have a chance to run for president, but that if I ever let Israel down, God would never forgive me. I will never let Israel down."[34]

One biographer notes that by this point in their relationship Clinton had come to see Vaught as something of a father figure.[35] Indeed, even if Clinton had initially picked Vaught's church for political purposes, he had ultimately come to view it as a fruitful choice—one that the governor himself might have later deemed providential.

Sunday School Teacher

As for Hillary, there was less second-guessing about her choice of church than about her faith in general. Joyce Milton notes that the religious practices of both Clintons were a subject for "image adjustment" in Arkansas, where voters were complaining that Hillary "had no religion" and that she likewise had not been regularly visible in church—some saw her as a godless liberal.[36]

Yet there were a number of signs that supported the legitimacy of Hillary's return to organized religion. Her openness about her Methodism was practiced with such frequency and gusto, while being completely consistent with her early church training, that it proved difficult to dismiss it as insincere and purely political. Throughout the 1980s, Arkansas's first lady traveled around the state giving a speech that explained why she was a Methodist—which, of course, did not hurt politically. In that stump speech, she spoke candidly about her faith, and specifically about her close theological ties to John Wesley, which had lain dormant for close to ten years:

As a member of the British Parliament, [Wesley] spoke out for the poor at a time when their lives were being transformed by far-reaching industrial and economic changes. He spent the rest of his life evangelizing among the same people he had spoken up for in Parliament. He preached a gospel of social justice, demanding as determinedly as ever that society do right by all its people. But he also preached a gospel of personal responsibility, asking every man and woman to take responsibility for their own lives . . . and cultivate the habits that would make them productive.[37]

These speeches were not just historical. Hillary also spoke of her personal relationship with the Almighty and explained how much of her beliefs was rooted in what she had learned at Park Ridge—experiences from her youth group days, which she shared with these Southerners. She also typically reiterated her favorite Wesley proverb: "Do all the good you can, by all the means you can, in all the ways you can, in all the places you can, at all the times you can, to all the people you can, as long as you ever can."[38]

Ed Matthews, pastor at the First United Methodist Church in Little Rock, recalled these speeches in an interview:

She was frequently out speaking in churches . . . particularly in Methodist churches. She has a lesson, a presentation, on "Why I am a United Methodist." It's not exclusive, not trying to exclude any other religions. She has a profound understanding of the Wesleyan tradition of grace. A profound understanding that we are all made by God in our own way with our own gifts to make our contribution and that the world is a better place because we've been here. That is not arrogant—it's a grace filled position. And she is a lawyer who herself has been quite well gifted.[39]

This resurgent evangelistic spirit was reminiscent of the Park Ridge Hillary, and of an adult who looked to be coming back to the faith. Reverend Matthews happened upon a telling insight into Hillary's faith at this juncture: "One of her favorite thoughts," he said later, "was that the goal of life is to restore what has been lost, to find oneness with God, and until we find this we are lonely."[40] This may have explained Hillary's separation in the 1970s, and her return to the flock in the 1980s. She was now in the South in the 1980s, a fish out of water, lost and lonely, and through her faith, she could restore what she had lost in the Northeast at Yale and Wellesley, again finding oneness with God.

While Hillary may not have been using her faith for political positioning, no doubt her political ideology was borrowing heavily from her faith during this spiritual resurgence. What Don Jones had given Hillary was a brand of Methodism fully applicable to her ideology; he had helped Hillary politicize her faith, which was now a fundamental source for her left-leaning ideas.

A barometer of Hillary's genuine return to the faith was the talks she gave outside of camera shot or reporters' notepads. According to Ed Matthews and others in the congregation, Hillary was a "regular" teacher in one of the Sunday school classes, alternating with other teachers. She was part of a so-called forum class—"forum" meaning that the weekly class had revolving leaders. She was part of the rotation. Hers was called the "Bowen class," after a man named Bill Bowen who was the acting administrative governor when Bill was running for president, and a prominent banker and lawyer in Little Rock. She was also an occasional guest teacher in other adult and youth classes, according to Pastor Matthews.[41]

One of the attendees, Willard Lewis, recalled Hillary teaching the Bowen class "several" times while also attending "several" times when the class was led by others, but not on a consistent basis.[42] According to Lewis, everyone in the group recognized that Hillary's regular attendance was prohibited by the fact that "she was a very busy lady," being both first lady of Arkansas and a lawyer with the Rose Law Firm, and that her attendance was "limited by events probably beyond her control, not by her choice." Lewis said that class members "always looked forward" to those occasions when Hillary could teach "She never failed to impress us with her verbal skills, her seemingly inexhaustible supply of information on just about any subject, and her intelligence, which I personally regarded as superior to that of her very intelligent husband."[43]

Lewis, a journalist who spent thirty-three years with the now defunct *Arkansas Gazette*, was on a first-name basis with Hillary. Though today he regrets that his "seventy-six-year-old brain" can no longer recall specific subjects she chose to address, his general recollection is that they were mostly of a religious nature and not having to do with politics or nonreligious themes, though she could veer into related subjects of interest. Says Lewis: "It seemed to us that even on short notice, Hillary was quite capable of literate, intelligent, and seemingly well organized treatises on just about any subject."[44] He also regrets that he cannot recall details, like whether Hillary opened or closed the class with a prayer.[45] Once, in a display that might set off alarm bells among today's more secular-minded Democrats, Hillary went so far as to invite the adult Sunday school class for a picnic on the back lawn of the governor's mansion.[46]

Lewis's short memory is unfortunate, since it seems that a Sunday school class led by Hillary would be more expected to focus on Methodist history, the teachings of Wesley, and social justice activism than some type of exegesis of Scripture. One would expect Hillary to select the topic for her Sunday turn at the lectern rather than picking up with a spot in the rotation where the last leader left off, with, say, John 6:53.[47] In other words, if Hillary had freelanced more than usual, jumping into the rotation with her own interpretation of a meaningful section of, say, 1 Corinthians or Romans or Hebrews or Galatians—rather than giving a vague explanation of why she was a Methodist or why Jesus loves the poor—it might tell us much more about the maturity of her faith and her spiritual path at that point.

When asked if there was any unfriendliness toward Hillary around the church because of her progressive views or her initial refusal to change her name, Lewis added: "I recall that there was some resentment in some more conservative quarters of the state . . . but I am not aware of any such sentiment in the church itself; quite the contrary, in fact. I think Hillary was widely admired by fellow church members and certainly faced no negativity in terms of [the] attendance or membership." Lewis notes that this was not a surprise because the First United Methodist Church "is, on the whole, a very liberal congregation."

But while Hillary herself was fairly well regarded, some churchgoers at First United were quietly beginning to have doubts about her husband, and as the 1980s progressed, rumors of Bill's philandering began to swirl around the church and the community of Little Rock. When asked, Lewis added that "except in the most general terms and then based solely on rumor and innuendo, Bill's peccadilloes were not widely known within the church itself."[48]

Nevertheless, they were becoming known to Hillary, and it would not take long for them to change the entire dynamic of the Clintons' marriage and their relationship with God.

CHAPTER 6

Hillary's Causes and Bill's Demons

With the political plus of having convinced the electorate that he and his wife were churchgoing and observant Christians, Bill Clinton won gubernatorial races in 1982, 1984, and 1986, easily defeating Frank White in 1982 and 1986 and Woody Freeman in 1984. After Clinton's victory in 1986, the governor's term was wisely extended to four years, meaning that the governor would no longer be in constant campaign mode. Meanwhile, the partners in power understood that winning in 1990 could seal Bill's path to the presidency and would require a course of public moderation, even as Hillary privately remained grounded in politics that were staunchly on the left.

The first lady of Arkansas launched a public education campaign to highlight problems faced by modern teens. She singled out sexual content, stating that society was "bombarding kids with sexual messages—on television, in music, everywhere they turn." In a throwback to the Park Ridge Hillary of the 1950s, she said that both parents and churches were failing teenagers in not doing enough to help them just say no to sex. "Adults are not fulfilling their responsibility to talk to young people, about the future, about how they should view their lives, about self-discipline and other values they should have."[1]

She added, in an interesting confession, that she herself, or, more specifically, the late-1960s Hillary, had not done enough in this regard. She was not quite sure as to how and where parents and churches had "got off the track" and had apparently jettisoned moral instruction. "Adults don't feel comfortable telling their children not to do things," said Hillary as a mother in the 1980s, "or they don't know how to communicate that message effectively. I'm trying to." Then, in a line that drew applause, she stated emphatically, "It's not birth control, but self-control."[2]

That well-received phrase ended up as a common line in her speeches from the period, but it was an interesting if not unexpected path for her to take. For years, Hillary had purported to be strongly opposed to the Catholic Church's stance against birth control. Yet remarkably, this line might be interpreted as an example of her going along with the Catholic Church's doctrine, advocating that people rely on their own moral responsibility and the practice of abstinence rather than the use of birth control. Given her widely stated opinions on birth control, it was a curious statement, made all the more curious by the fact that listeners almost always responded with loud cheers of support. On the other hand, she might have meant that proper self-control is the real issue, not birth control. Unfortunately, she did not elaborate on what she meant.

The Governor's School

One of the most interesting discoveries concerning Hillary's activities in this period was dug up by biographer Joyce Milton, who unveiled what she called "one of the stranger footnotes to the Clintons' impact on education in Arkansas."[3] It was the so-called Governor's School, a six-week summer program for four hundred gifted high school seniors. Bill Clinton said that the program, started in 1979, was his "dream come true." As Milton notes, however, the summer school was the pet project of the governor's wife. Indeed, it had Hillary's fingerprints all over it, a total reflection of her life and interests.

Now nearing its thirtieth anniversary, the Arkansas Governor's School is hosted on the campus of Hendrix College, the Methodist institution in Conway, Arkansas. According to the program guide, the school is a residential program for "gifted and talented high school seniors" from all over Arkansas. There is no charge for tuition, room, board, or even classroom materials—the entire program is fully funded by the state of Arkansas.

Though it began during Bill's first term as governor, the school did not officially hit its philosophical stride until the mid-1980s. Joyce Milton speculates that the Governor's School had become a "reincarnation" of Don Jones's University of Life, or, at the least, Hillary envisioned it that way. The analogy seems apt since much of the Governor's School curriculum was more postmodern and left-leaning than a lot of what was available in the public school curriculums. Over the years, students would watch films like *Do the Right Thing* and *The Times of Harvey Milk*, read works such as Beckett's *Waiting for Godot* and Sartre's *The Flies*, and study the theories of sociologist Herbert Gans, author of the 1971 manifesto "The Uses of Poverty: The Poor Pay All," and the book *The War Against the Poor*, which states that "The poor are far more vulnerable, and racial minorities among them even more so."[4] "Poverty can be eliminated," wrote Gans in a 1971 work likely read by Governor's School pupils, "only when . . . the powerless can obtain enough power to change society."[5]

One Governor's School pupil whom Milton consulted with, Eddie Madden, attended in 1980, a year that featured an appearance by Hillary, whose theme, according to Madden, was to trust big government over big business, which would have been consistent with what they were learning in their studies of Gans. Madden said that another speaker, a physicist, claimed that a scientist, or, at least, a good scientist, could not be a religious believer, since science and religion, and faith and reason, were mutually incompatible. "The Book of Genesis should be read, 'In the beginning man created God,'" the speaker instructed the class, a charge that initiated an intense debate about the Book of Genesis, John's Gospel, and religion generally.[6]

Madden stated that another speaker at the Christian college during that summer of 1980 was a feminist named Hope Hartnuss, who he said lectured on the collapse of women, arguing that women had been fine until Christianity came along. "Christianity was anti-woman and anti-sexuality," said Hartnuss, according to Madden. The misogynistic church had persecuted the female gender because of its "fear and hatred of women."[7]

A decade later the same curriculum was still being followed and administered, only with different participants. Milton offered another example from 1991—the last year in which Mrs. Clinton could devote any time to the program—in which a feminist named Emily Culpepper reportedly fulminated against Christian patriarchy. The writing, says Milton, dismissed Christianity as "compost" and instructed the recent high school grads that belief in Christ's divinity was "no longer just implausible [but] offensive."[8]

Milton interviewed a 1991 participant named Chris Yarborough, who said that the goal of the program seemed to be to "deprogram" young people away from the traditional values they had learned and to inculcate them into the brave new world of postmodernism, with special attention to "feelings" and so-called critical thinking, a critical thinking that did not include exposure to alternative viewpoints.[9]

While it remains unclear precisely what personal role Hillary had in shaping the curriculum, the reality is that some of what went on there seems to have run counter to her religious revitalization of the 1980s. Though she felt comfortable going from church to church speaking publicly about her Methodist values and roots in an attempt to polish her churchgoing image, these religious beliefs were hardly on display during the years when she played a key role at the school.

On the contrary, it was her political ideology that took front and center stage here, an interesting choice given the impressionable audience in attendance. Whereas during her religious speeches, she was preaching for the most part to people who were already committed Christians, at the Governor's School there existed a clear opportunity to influence a generation of young minds—much in the way that Don Jones had shaped hers. The fact that this school, in which she played such a crucial role, chose the writings of Sartre instead of John Wesley was telling.

Don Jones's Disciple

The Governor's School demonstrated that the idea of Don Jones continued to influence her faith, and in actuality the real Don Jones was still a source of spiritual guidance for Hillary. One such instance occurred for Hillary when she found herself grappling with the issue of capital punishment. For several years, Bill had been supporting capital punishment as a Southern governor, but Hillary had long had spiritual doubts about the Christianity behind supporting such a policy.

While the topic had long provided Bill with a good issue to help position himself as a moderate, Hillary seemed more ambivalent. As Jones told Judith Warner, he discussed this specific issue with Hillary when Governor Clinton was once considering whether to commute a capital sentence for a serial killer and rapist. Hillary "agonized" over the decision, and consulted Jones. Jones told her, "Well, I believe there is such a thing as punitive justice; that's part of the whole concept of justice. And I think some people have forfeited their right to life because of the heinous deed that they've committed." In response, says Jones, Hillary told him, "Well, I think I agree with you."

However, says Jones, it was evident that Hillary "was struggling with the question of could she conscientiously as a Christian say that.

There was a tad of uncertainty about that. And I attribute that to her faith."[10]

This was not the only time that she leaned on Don Jones during her time in the Arkansas governor's mansion. In April 1987, she had some business to conduct in New Jersey on behalf of the Rose Law Firm, and so she called Don Jones, then at Drew University, to see if they could get together and chat over coffee at Newark Airport. As it turned out, Jones had class that evening, and instead they arranged for Jones to pick her up at the airport and take her to Drew, where she could speak to his business ethics class.[11]

Hillary seemed an odd choice for this lecture, as she had never owned, managed, or even worked for a business outside of a law firm; her entire career had been in academia, law, and nonprofits. Nonetheless, Jones liked the message that she was eager to deliver. At Drew, Hillary attacked materialism, selfishness, and excessive individualism, rooting her worldview in Don Jones's Methodism. "We are experiencing a crisis of meaning and a spiritual crisis," said Mrs. Clinton. She spoke of "the hurt, emptiness, confusion, and loss of meaning that characterizes much of our society."

Caricaturing the Reagan boom of the 1980s as a malevolent "Decade of Greed," she said that corporate America was the culprit for this era of excess. The corporate world's obsession with short-term profit at the expense of morality and ethics, she said to her former teacher's students, was subverting America's moral, spiritual, and democratic values, and the family itself.[12] Once again, she was delivering the social justice gospel of the religious left, focused on economics and the dangers of capitalism, exempting hot-button social-moral issues like pornography, promiscuity, marital infidelity, divorce, and abortion.

Bill's Demons

Among those touchy contemporary social-moral issues that Mrs. Clinton avoided at Drew University in New York were two she could not escape in Arkansas—promiscuity and marital infidelity. Hillary had faced challenges growing up under the roof of Hugh Rodham, but nothing like those she faced with the man of the house at her home in Little Rock. And those problems, as much as the Clintons' loyal supporters to this day prefer not to discuss them, are a central part of the story of Hillary's faith.

It is difficult to say when, exactly, Bill Clinton began his extramarital affairs. One sensational recent account goes so far as to say that his infidelity began immediately, claiming that during the Clinton wedding reception that followed at the home of friend Ann Henry, a guest caught Bill kissing another woman in one of the bathrooms.[13] Whether this truly happened is something that this book cannot prove, but the story indicates the kinds of claims and anecdotes that would follow Hillary and Bill literally from the very start of their marriage.

If Bill was not cheating on Hillary by October 12, 1975, day two of their marriage, he was doing so by the 1980s. Hillary had suspicions, and rumors were rampant. Despite the accusations that swirled around Bill, there was little sense of how Hillary reacted to the situation, and how her faith was impacted by Bill's behavior. To this day, the mystery surrounding Hillary's reaction to her husband's behavior has swelled, becoming one of the great public questions of the couple's marriage. Over the years, many sources have reported that Hillary was deeply troubled by these infidelities, and she took her turmoil to God, or at least to a man of God—a minister. This would be said in the 1980s and again in the 1990s. Those who say she took it to God, or to a pastor, usually portray her sympathetically, tragically, whereas others see her as that "partner in power" who expected this behavior from her husband and looked past it, always with her gaze fixed upon seats of political power—a trade-off.

The one source who can shed some light, if not answer many of these questions, is Hillary's pastor at the First United Methodist Church in Little Rock, Ed Matthews, who claims that Hillary did not seek him out for counseling in the 1980s. That being said, Matthews contradicted himself on certain points over the course of several different interviews, so it is possible that his version of events may be incomplete.

When interviewed in October 2005, Matthews volunteered information regarding charges of Bill Clinton's womanizing, without prompting. He said: "There was a false, fabricated story on when Clinton was governor . . . talking about how much time I spent [counseling] them because of his so-called womanizing. There wasn't one word of truth to that."[14] Asked if Hillary ever approached him in the 1980s when the rumors about her husband's affairs began to spread, Matthews said, "She never approached me in any way."[15]

Importantly, Matthews says that while the couple did not come to him for counseling, Hillary was very much in personal crisis, suffering a broken heart, and sought solace in the Book of Psalms. "I think she especially appreciated the Psalmist's ability to be honest," said the senior minister. "In the Psalms you can find extreme mood changes in any one of the chapters. The Psalmist could be mad at God—*How could you let this happen? Why is it this way?*—and then wind up asking forgiveness for doubting [God]. I think that kind of honesty and straightforwardness, as well as the mood changes that any one Book of the Psalms could carry comforted Hillary. There is also a sweetness and tenderness about the Psalms that help when your heart is really hurting."[16]

Matthews also had insight into what Hillary might have said to Bill at the time, telling Gail Sheehy: "I have a strong feeling that Hillary would have been confrontational with her husband on these matters. At that time, the more respected friends of the Clintons were anxious that he and so-and-so had had an affair, and that Hillary was now aware of it."[17] According to Sheehy, Hillary never mentioned names to her pastor, but discussed confronting her husband. Matthews told Sheehy, "Clinton admitted it, and at that point maybe they did get some counseling."[18]

Despite Matthews's claims that he was not involved in counseling the couple, various biographers over the years say that he did. Norman King's book on Mrs. Clinton says that at one point during 1988–1990, Mrs. Clinton became so engrossed and fed up with the rumors about her husband that she contacted Matthews. "The pastor immediately agreed to meet with the Clintons for counseling," says the book. "Now it was up to Hillary to convince her husband that the meeting was necessary." King continues the narrative, describing Hillary's accusation to her husband that he was having affairs, and his agreement—after an argument with Hillary—to seek help:

Clinton finally agreed to meet with Dr. Matthews and Hillary. It was just a few days before Christmas when they assembled in the pastor's study at the church. "The meeting was charged with emotion," said one unnamed source. "Hillary said that if Bill would change, she'd try to make the marriage work."

Dr. Matthews asked Clinton, "And what about you, Bill? Are you ready to save this marriage?"

He said that he would do anything to save the marriage. "I love Hillary and Chelsea more than anything in the world."

A close friend of the family recalled, "Hillary and Bill wept as they held hands, knelt, and prayed together. Bill gave Hillary a solemn promise to change his ways, and they renewed their pledge of love for each other."

There were several more sessions in Dr. Matthews's study, and in the end the two seemed closer than they had been before.[19]

This touching account of forgiveness and reconciliation, published in 1993, seems to have been picked up and retold by others, including biographers who do not cite sources, making it unclear where their information came from.[20]

A 1996 book by Martin Walker says nearly the exact same thing, without citing a source. Walker also says the Clintons went to Matthews for counseling. "Under Reverend Matthews's direction," writes Walker, "they held hands and knelt to pray together" in Matthews's study, "and Clinton promised to change his ways, to work harder at being a better husband and father, and to devote more time to his family. There were repeated sessions in the pastor's study, and Hillary too pledged to change."[21]

Walker added more beyond what the 1993 book reported: He says that Hillary went on a "ferocious diet," losing twenty pounds, and changed her hairstyle and even began to dye her hair blond. He says she spent $10,000 on a new wardrobe, including $2,400 for a cashmere jacket, and went to a beauty salon for advice on cosmetics. "The marriage had been through a severe crisis," wrote Walker, "and their friends were heartened to see it restored in the course of 1990. They had reaffirmed their marriage vows before Reverend Matthews."[22]

Regardless of Matthews's disputed involvement, another debated element of the marriage's problems during this time is when the situation was formally resolved. Sheehy quotes Betsey Wright, one of Clinton's staff members who dealt with the political fallout over Bill's philandering more than any other aide: "In the end," said Wright, "they made a commitment to work on and save their marriage." This must have happened, as Sheehy intimates, "at some point" early in 1990, when, according to Sheehy, Hillary sent Bill to a (undated and undocumented) "Come-to-Jesus meeting," which in the South means a day of reckoning when a sinner confesses his sins and asks to be forgiven.[23] Sheehy says that Clinton went to that meeting, and vowed that this was "the last time," and that he would "start over" and straighten out. He wanted to be faithful, recorded Sheehy, and wanted forgiveness.[24]

Even as these stories of reconciliation have circulated over the years, there have also been persistent rumors that the Clintons briefly considered divorce. Judith Warner reports interesting information on that front, noting that Hillary was furious that her partner in power was jeopardizing their mutual political greatness because of his inability to control his sexual appetite—even writing off the 1988 presidential race: "Infidelity seems to have been the prime issue," reported Warner. "Bill Clinton's inability to run for president in 1988 because of his personal problems was just too infuriating, Hillary told her friends. What was she supposed to do with the rest of her life? Serve tea in the Arkansas governor's mansion?"

According to Warner, Bill—always eager to share his feelings with the world—even broached the subject of divorce in the intimate setting of the National Governors' Association, adding: "Political considerations clearly were never far from his mind." But in the end, writes Warner, "faith—some kind of faith—in the marriage prevailed."[25]

For whatever reason, Hillary reportedly took divorce off the table. Gail Sheehy quotes Betsey Wright, who was adamant: "It [divorce] absolutely was not an alternative that she gave him."[26]

Here, too, there are conflicting accounts. Martin Walker, for example, claims just the opposite, stating that Hillary had laid down the law, telling Bill: "Unless you're ready to change, we're getting a divorce."[27]

Whatever Hillary said about divorce, and whoever counseled the couple, the reported reconciliation made little difference to Bill, who soon went right back to the dysfunctional behavior. Joyce Milton writes that Clinton's pursuit of other women, "if it was interrupted at all, soon resumed." In turn, says Milton, Hillary, "as she had before," now once again "expressed her determination to make the marriage work by dedicating herself ever more fiercely to the business of getting Bill elected." Adds Milton: "This dynamic, so obvious to outsiders, inevitably led to speculation that the Clintons had a marriage of convenience, based solely on their shared political ambitions." Yet, says Milton, "It wasn't that simple. There was very little that could be called 'convenient' about the Clintons' marriage. But Hillary at this point had invested almost as much effort in Bill's career as he had, a consideration that had to weigh heavily against any thought of divorce."[28]

The Chelsea Factor

Not to be forgotten amid these marital problems was a harsh reality caught right smack in the middle of this: a sweet little girl named Chelsea. By the late 1990s, it was hard for her to ignore that something was wrong. With Arkansas's small size, the Clintons had become something of a soap opera, and the rumors were unavoidable. To cope, Chelsea turned to her church—the same place where Hugh Rodham's daughter and Virginia Clinton's son had once found comfort at the same age when the adults in their lives misbehaved.

Several years prior, Chelsea had joined her mom's church, where she was baptized and confirmed. But that did not happen right away: According to Judith Warner, Hillary felt it was important for Chelsea to choose her own denomination, believing that by doing so Chelsea's confirmation would take on added significance for the young girl. Though Hillary did not have the luxury of making her own denominational choice, she relished the remembrance and the significance of her confirmation at First United in Park Ridge and wanted her daughter to feel the same spiritual connection to whatever church she chose. Around the age of ten, Chelsea went to her father's church for a time, attending Sunday school at Immanuel Baptist. Later on, she began to visit Hillary's church, and shortly thereafter decided that the Methodist church was the right choice for her. Hillary was reportedly relieved when Chelsea decided on Methodism, as was Dorothy Rodham, who, during those days of Chelsea's Baptist Sunday school, had once heard the little girl bark out a charge of blasphemy when the words "Oh, my God!" passed through her grandma's lips. That one scolding from the little girl was all it took. "Dorothy Rodham didn't relish the idea of a fundamentalist Southern Baptist grandchild," says Warner.[29]

Chelsea felt more at home at First United Methodist in Little Rock, her mom's church on Center Street. On the day she was confirmed at the church, Bill Clinton joined his wife for the ceremony. Senior Pastor Ed Matthews had many dealings with the Clintons simply through Chelsea and the process of her confirmation. He worked with Chelsea for nine months, during which he says he "spent a good bit of time with each of the parents," and felt "greatly inspired by all three of them." Matthews said that while Bill Clinton attended Immanuel Baptist, "when anything involved Chelsea at our church—she was always in the Christmas pageants and the children's choir—he was always there in the front row cheering her on." He added that Chelsea was "greatly devoted" both to her parents and to her confirmation classes.[30]

Matthews says that he saw a need to offer some counsel to Chelsea when the sexual accusations were being made against her father. He remembered:

There was one particular Sunday, when we all came out after the service, and there was a black man, whose name was in the press, who was outside that day, and was very rough on Governor Clinton and would make accusations about his womanizing, and this was before the accusations were very public. We all had tracts on our windshields. Absolutely the most vulgar thing you ever saw in your life, accusing the governor. It was then that I thought, *What do they tell Chelsea?*

So, I called Hillary that afternoon, and this was one of the few times I initiated a conversation about anything like this, and asked her if they were preparing Chelsea. She told me that, "We do talk to Chelsea about this. We do sit down with her, we recognize the public life that we live, that these kind of things come along and [we] want to prepare her for it." She was saying, in essence, we know the truth about these matters and we don't try to fight and be defensive. But we know the truth and live the truth.[31]

"As it turns out," added Matthews, "some of these things were true, but this was her effort to prepare Chelsea."[32]

One of the more dubious ways that Chelsea's mom and dad prepared her was shared by Mrs. Clinton in her book *It Takes a Village*. As Bill prepared for another election, Hillary explained to Chelsea that Daddy had enemies, mean men who said cruel, untrue things about her father. Hillary recommended a game: Chelsea could pretend to be her daddy, campaigning and telling people that she was the nice governor, Bill Clinton, who had "done a good job" and "helped a lot of people." "Please vote for me," she said.[33]

Just then, Hillary and her husband descended, morphing into the *evil* men that campaigned against Bill, snarling, saying "terrible" things about Chelsea's daddy, about how he was a "really mean man" who hurt people. When the frightened little girl's eyes filled with tears, and she protested, "Why would anybody say things like that?" the Clintons backed off, mission accomplished—for now.

The child care champion judged the role-playing game effective and continued the systematized training for the weeks ahead over several more dinners, until Chelsea had "gradually gained mastery" of her emotions and was equipped to "discern motives" of her father's accusers—which apparently meant that she was taught to reflexively assess negatives tossed at her father as mean-spirited fibs.[34] Surprisingly, Hillary wrote openly of her unorthodox method, not grasping that some would later question it as an ill-advised form of borderline psychological child abuse.[35]

Whether it was out of their love for each other, their love for Chelsea, or their love for politics, the Clintons' marriage managed to survive the turmoil of the late 1980s and early 1990s. Despite the discrepancies over the extent of Ed Matthews's personal involvement in the eventual reconciliation, it seems clear that, in a variety of ways, Hillary leaned on her faith to help her during this troubled time. Spending time with Chelsea in confirmation class and being a presence at her church in Little Rock, Hillary more than once looked to her faith, and in the midst of the repeated accusations and public controversy, she appears to have turned to God to give her the strength to hold the relationship together.

But although her faith in God during this period seems undeniable, questions persist about her ultimate motivations for remaining by her husband's side, and these questions would only get louder as the public stage of their lives got bigger. While it is safe to speculate that as a wife and as a mother she wanted to keep the family together, the debate continues to rage over why during these difficult years she was so determined to stand by her husband, who seemed so incapable of changing his ways. Indeed, this is a question that people have been asking about Hillary for years: As an independent, intelligent woman whose politics and social causes were very much a product of the women's liberation movement, why would she stay with a man who repeatedly defied her requests to change?

While the Methodist denomination (and all denominations, for that matter) clearly encourages a happy marriage as the proper environment to raise a child, it seems uncertain that faith and family were enough to hold the Clintons together. More plausible, especially given her actions in recent years, is that Hillary's own political ambition played a part in her ability to turn the other cheek to Bill's indiscretions.

This difficult time and the host of questions that it elicited would only grow in legend as the profile of the couple increased, and though the couple would always share God, it was unclear whether faith or political ambition would be the glue that would hold them together in the turbulent years that lay ahead.

CHAPTER 7

Taking Power

In November 1990, forty-four-year-old Governor Bill Clinton won his fourth consecutive election campaign, and fifth overall, trouncing the Republican challenger Nelson Sheffield, 57 percent to 42 percent. With a fifth term locked up, the path to the promised land—the White House—was wide open. But with this ambition also came a drawback: The Clintons would be open game to the national press, including relentless critics on the right, eager to dig up dirt on the governor. Here, Bill had been his own worst enemy, as there was no shortage of mud to clog the political effort to make the governor shiny and palatable to the American public.

Still, by the summer of 1992, the once unthinkable seemed possible: Republican president George H. W. Bush, who only a year earlier had a 91 percent approval rating because of his success in the Gulf War, was suddenly trailing this no-name governor from one of the smallest states in the Union. Most Democrats had dropped out of the race in 1991, not wanting to waste their time challenging such a popular incumbent. Clinton, however, had kept his hat in the ring, with stunning results.

Just then, one of those women from the past, a bleached-blond TV news reporter from Little Rock named Gennifer Flowers, came out of the woodwork to tell the world about the man she slept with throughout the 1980s—behind the back of Mrs. Clinton. She even had taped phone conversations with the governor. She called a press conference, and a media sympathetic to Bill Clinton could do nothing to stop the avalanche that followed.

Hillary, however, could do something—she was not about to watch all of this get blown up now. She agreed to stand by her man, sitting next to him for an exclusive interview on CBS's *60 Minutes*, watched by millions, and performed a kind of miracle that would have made Jimmy Swaggart swagger. The two conceded some difficulties in their marriage in the past, but, they added, they were in love, and beyond those problems. When reporter Steve Kroft asked the Clintons if they had some kind of an "arrangement," the two laughed heartily, and handled the question with aplomb.

Enough of the voting public accepted the Clintons' word; millions of Americans were willing to look beyond the past and vote for Bill, judging that the good governor's peccadilloes were behind him. His wife had forgiven him, after all. Besides, if elected president, he surely would not engage in this kind of activity in the White House.

What about Hillary? Was she a victim or a cold political operative in this situation? Here again, the sources differ. Norman King says that to cope with the betrayal by her husband, not to mention the national humiliation, Hillary "would make a habit, even when on the campaign trail," of carrying a "tiny little Bible that has Proverbs, Psalms, and the New Testament."[1]

King did not name his source, but it may have been Ed Matthews, who said in an interview for this book that during the 1992 presidential campaign, "She carried, in her purse, a Bible, and specifically the book of Psalms." However, when asked in a follow-up interview to expand on this imagery of Mrs. Clinton carrying a copy of the Book of Psalms with her in her purse, an image that might jolt some of her detractors, Matthews retracted: "I wouldn't say that she carried it; it was her companion that she referred to regularly. In her purse all the time would be an overstatement." Matthews said she carried the Psalms with her in heart and mind.[2]

Winning the White House

Of course, Bill's partner could not claim that his actions alone were a liability to their joint political ambitions (and their marriage). The woman that Bill hurt with his sexual past was hurting the Clinton-Gore ticket with her political past.

The Clinton strategy for 1992 was to run Bill as a "New Democrat," a moderate Democrat. This was well planned: From 1990 to 1991, Bill chaired an important group called the Democratic Leadership Council, a collection of Democrats who understood that if their party was ever again to win the White House, they would need to stop running ultraliberals at the top of the ticket—no more McGoverns or Mondales. There, he was joined by a onetime moderate, the pro-life senator from Tennessee, Al Gore.

A moderate Democrat must, of course, be a religious Democrat. And Bill Clinton noted during the campaign that he was such a Democrat. "I pray virtually every day, usually at night, and I read the Bible every week," the candidate told *U.S. News & World Report*. He added that he believed strongly in "old-fashioned things" like the "constancy of sin, the possibility of forgiveness, the reality of redemption."[3] While these words might sound hollow from some candidates, from Clinton they were in line with his long-established beliefs and life experiences.

Yet, in trying to run as a New Democrat, Bill Clinton's biggest obstacle was not his background but that of his wife. He strove to cultivate an image of a sensible Southern governor, not a "Massachusetts liberal" like Michael Dukakis, the Democratic nominee against Bush in 1988. However, his wife had actively supported many left-leaning causes that were far from moderate, making it difficult to convince much of the public that she was as middle ground as her husband. With Hillary around, it seemed as though there was a Massachusetts liberal, or at least a Wellesley, Massachusetts, liberal, on the ticket after all. This forced Bill into a defensive and disingenuous denial, arguing that to portray his wife "as some sort of left-wing figure based on her activities over the past 10 or 15 years is patently absurd."[4]

In the end, it would not matter. Clinton and his running mate spoiled George H. W. Bush's bid for reelection. Clinton did not receive a plurality of votes, and was helped enormously by the entry of Texas billionaire and Bush rival H. Ross Perot, who walked off with 20 percent of the vote, one of the biggest takes by a third-party candidate since Teddy Roosevelt in the 1912 election. Clinton needed only 44 percent of the popular vote to win handily. Meanwhile, Bush received only 36 percent, an astonishing drop from his 91 percent Gallup approval rating only a year earlier during the Gulf War—a plummet that his vice president, Dan Quayle, called the single greatest political free-fall in American history.

Clinton and Gore had tremendous success arguing that the economy was at its lowest point since the Great Depression, an exaggeration maybe not outdone by any politician since the Great Depression. Rather, the 1990–1991 recession, in complete recovery and full rebound by the time of the November 1992 vote, was America's worst since the 1982–1983 recession of the first Reagan term. No matter. The gross, unchecked exaggeration did wonders at the ballot box.

Hillary's husband defeated George W. Bush's father. The junior Bush was certain that the more honorable man had lost; he hated "to see a good man get whipped."[5]

The First Lady–Elect

Only two or three weeks after the November triumph, Hillary took to the dais to thank Marian Wright Edelman, and to begin looking for the next cause to advance in the left's sweeping social agenda. It was fitting that the first lady–elect's first major postelection speech was at a CDF dinner in Washington, where she described Edelman as her "mentor and leader" and asked a question that immediately signaled how those on her side would caricature conservative Republicans who stood in the way: "What on earth could be more important than making sure every child has the chance to be born healthy, to receive immunizations and health care as that child grows to be stimulated and learn so a child can be ready for school?"[6] Here again, CDF could be that vehicle for social justice. And Hillary Rodham Clinton knew exactly where to drive it next: national health care.

As the presidential inauguration approached, Hillary's husband, that Southern boy she met at Yale, was ready for the top job in the world. The partnership had persevered and produced amazing results; the possibilities seemed limitless. And though the perception of the Clintons at this time—by the right and the left—was not one that called to mind church bells, there was a man, then unknown to the national press, who stepped forward to tell Americans about a side of Hillary they had not seen. "This may sound corny, but the key to understanding Hillary is her spiritual center," a fellow named Don Jones explained to *People* magazine as the inaugural neared. "Unlike some people who at a particular age land on a cause and become concerned, with Hillary I think of a continuous textured development. Her social concern and her political thought rest on a spiritual foundation."[7]

Jones's words were a revelation to everyone but him. Indeed there was far more to the new first lady than the public knew.

Bill Clinton, President

On the morning of January 20, 1993, Hillary and her husband readied to move into the most powerful house in the world—the White House.

That morning, in the final activity before the inaugural ceremony, a prayer service was held by the Clintons at the Metropolitan African Methodist Episcopal Church. The service, said Bill, "was important to me." He picked the participating clergy, the music, and the vocalists, with input from Hillary and Al Gore. Both of the Clintons' ministers from Arkansas participated in the service, a Baptist and a Methodist, as did the pastor of Al and Tipper Gore, the father of close aide George Stephanopoulos, and the Greek Orthodox dean of the Holy Trinity Cathedral in New York.[8]

Later that afternoon, Hillary's husband was inaugurated the forty-second president of the United States. The son of Virginia, that hardworking nurse from Arkansas, had gone from rags to riches. That poor fellow Billie Blythe, who had drowned in a few inches of water off Highway 60 on his way back to Hope, could not have imagined in his last breaths that he was leaving quite a legacy in his wife's womb: a future president. On the other side of the marriage, the girl from Park Ridge had now fulfilled the wildest expectations of those young women from Wellesley: She had made it to the White House. Of course, she was not president, but she was married to one, and would with him share power to an unprecedented degree for a first lady—the most powerful since Eleanor Roosevelt.

Few Americans then imagined that Hillary Rodham Clinton planned to one day parlay that seat near the presidential inaugural stand into her own presidential bid. To do that, she would, like her husband, require the votes of moderates, especially religious-minded voters. And yet, from day one, Bill Clinton took an action strongly favored by his wife that was poised to separate them from those voters.

In a flurry of unprecedented Oval Office activity, on that January 20, with Hillary's full backing, Bill Clinton signed five executive orders dramatically increasing the federal government's support and funding of elective abortion. He thrilled his wife and other supporters of abortion rights, yet he also made instant enemies, including the very religious constituencies that his wife would one day doggedly pursue in her bid to have her own day at that same inaugural stand with her hand on the Bible.

For instance, four days after the signing of the executive orders, the Vatican newspaper, *L'Osservatore Romano*, ran a dire editorial, stating grimly that the "renewal" that the man from Hope had promised during his campaign would now come "by way of death" and "by way of violence against innocent human beings." The stage was set. As papal biographer George Weigel noted, it was the opening salvo in what would become "the most serious confrontation ever" between the U.S. government and the Holy See.[9]

A New Methodist

That battle over abortion, however, was to come. For now, and now that they were in Washington together, and shared the White House together, the Clintons at last decided to join the same church. As George W. Bush later did for his wife and first lady, and Bush's father had done for the most recent first lady, Bill Clinton likewise joined the denomination of his wife and started attending a Methodist church. The Clintons chose Foundry United Methodist Church, less than a mile from the White House. This was a good match for the Clintons, a mixed, self-prided "inclusive" congregation. Young singles and married couples typically sat up front on the far left, whereas gay congregants preferred the far right.[10] Said Bill, listing the reasons he and Hillary picked Foundry: "We liked Foundry's pastor, Philip Wogaman, and the fact that the church included people of various races, cultures, incomes, and political affiliations, and openly welcomed gays."[11]

The Reverend Dr. J. Philip Wogaman is president of the American Theological Society and professor emeritus of Christian ethics at Wesley Theological Seminary. He is well known in religious, political, and academic circles, and particularly among Methodists. He was a delegate to the denomination's General Conferences of 1988, 1992, 1996, and 2000, as well as a member of the World Methodist Council from 1986 to 1991. His published works include seventeen books. From 1966 to 1992, he not only taught at Wesley Theological Seminary, but served as dean from 1972 to 1983. Beginning in 1992, he became pastor at Foundry, a large historic church in the heart of the nation's capital, where he stayed until 2002, meaning he was there for the entire Clinton presidency. In this capacity, he would spend more time giving spiritual counseling to the Baptist rather than the Methodist in the Clinton family.

As a liberal, Wogaman could be safely expected to focus on social justice and not cause the Clintons political embarrassment by doing untoward things like lecturing them from the pulpit on inconvenient matters like the sanctity and dignity of human life. In this respect, he was a perfect fit, particularly as it pertained to Hillary's brand of pro-choice Methodism. Wogaman himself stated that he aimed to be "prophetic" in his sermons "without embarrassing the president."[12] Indeed, the staunchly pro-choice Wogaman even opened his pulpit to fellow Methodist and author of *Roe v. Wade*, Harry Blackman, one day in 1995. According to Mark Tooley of the Institution on Religion and Democracy, who was seated among the congregation, Wogaman was visibly displeased when Blackman, who was scheduled to visit, canceled his visit at the last minute because of pro-life demonstrators outside of the church. The reverend offered a stern rebuke to the demonstrators, calling their action a "tragedy."

According to Wogaman, the Clintons attended the church on a regular, consistent basis from their first weeks in Washington, and ultimately through the life of the administration, and generally sat three rows down from the front off the right center aisle. By regular attendance, Wogaman meant "more than once a month, several weeks in a row," noting that as president, Clinton was often out of town, and could not attend every week, even if he so desired. Yet, said Wogaman, "When in town, it was always assumed he would be there." This, he says, was true for Hillary as well.[13] They generally did not miss church.

When interviewed for this book, Wogaman had more to say about Bill's faith than Hillary's, since Bill kept him busier. "I was well acquainted with Hillary," says the pastor, "but did not get involved with the same sort of spiritual counseling."[14] He called Hillary "a born and bred United Methodist, and as far as I know has been faithful to that commitment all her life." He noted that during her years in the White House, she formally remained a member of the First United Methodist Church in Little Rock. Wogaman added that while at Foundry, both Bill and Hillary met "periodically" with a parent group. This group was made up of the parents of children in the youth group, which included Chelsea, a regular participant.[15] During one instance in early 1996, the first couple went to a meeting of the teen group in which the adolescents openly discussed what bothered them about their relationships with their parents. Said Hillary: "[I]t helped to have another child say what your child didn't want to say to you directly"[16]—and presumably in front of the president and first lady.

The Politics of Meaning

At Foundry, Hillary was engaged in conventional (albeit left-leaning) Christianity, standard modern Methodism. Yet convention was put aside as she developed a keen interest outside of Methodism, a fusing of her political and spiritual worldviews, through something called the "politics of meaning." It was a search for truth partly inspired by a personal crisis involving the first man in her life—her father.

In 1987, Hugh and Dorothy had moved to Little Rock to be closer to their daughter—for Hugh, physically so, if not emotionally. From there, they watched their daughter and son-in-law slug their way to the Oval Office. Of course, this was not about to turn Hugh into a Democrat, and the lifelong Republican remained loyal to the GOP.

In March 1993, only weeks into the Clinton presidency, Hugh suffered a massive stroke and slipped into a coma. Hillary was obviously upset, particularly because a coma leading to death would never allow for a meaningful good-bye. The event proved a stark reminder of how difficult life can be: One minute, she was pinching herself to savor the accomplishment of earning a seat near the pinnacle of power; then came this. It was surely a reminder of what mattered most: family and faith, God and the eternal.

She thought of Hugh's faith in God, vividly recalling how he had "said his prayers kneeling by his bed every night of his life, until he had [his] stroke."[17] Over the next three weeks, she and her mother and brothers prepared for the inevitable, and Hillary gingerly got on with the business of being first lady.

That business included a major April 6, 1993, address at the University of Texas at Austin, as part of the college's annual Liz Carpenter Lecture Series. In this speech Hillary introduced a phrase to the broader public: "the politics of meaning." The audience was expecting Mrs. Clinton to talk about health care. She had been assigned head of the president's Task Force on National Health Care Reform—an extraordinary policy responsibility for any public official, particularly a first lady. That day at the University of Texas, she spoke about health care, all right, but much, much more.

"[W]e have to begin realistically to take stock of where we are," she began, "to be able to understand where we are in history at this point and what our real challenges happen to be." This, she said, was not merely an American problem, but a North American problem, a European problem, a Western problem. All were facing "the rumblings of discontent, almost regardless of political systems, as we come face to face with the problems that the modern age has dealt with." According to Mrs. Clinton, the modern problem was this:

Why is it in a country as economically wealthy as we are despite our economic problems, in a country that is the longest-surviving democracy, there is this undercurrent of discontent—this sense that somehow economic growth and prosperity, political democracy and freedom are not enough? That we collectively lack, at some core level, meaning in our individual lives and meaning collectively? . . . And it isn't very far below the surface because we can see popping through the surface the signs of alienation and despair and hopelessness that are all too common and cannot be ignored.

Mrs. Clinton was speaking for herself, but she was also speaking for a large number of disaffected liberals who saw the 1980s as Ronald Reagan's "Decade of Greed," which had failed to personally inspire them, especially those for whom spiritual faith was not a factor in their lives, or no longer a factor. She continued: "We are, I think, in a crisis of meaning . . . . What does it mean in today's world to pursue not only vocations, to be part of institutions, but to be human?"

She cited the recent death of the Machiavellian Republican political strategist Lee Atwater, who died prematurely of cancer. In those final weeks of weakness, Atwater had issued a number of mea culpas and spoke of the need for "a little heart, a lot of brotherhood," that he said was missing in society. Her citing of Atwater was lambasted by some conservatives as a form of political exploitation of a dying man, which was not fair to Mrs. Clinton, who merely endorsed and expanded upon what Atwater had said. Atwater had also spoken of a "spiritual vacuum," which prompted the first lady to ask:

"Who will lead us out of this spiritual vacuum?"—this answer is "all of us." Because remolding society does not depend on just changing government, on just reinventing our institutions to be more in tune with present realities. It requires each of us to play our part in redefining what our lives are and what they should be.

We are caught between two great political forces. On the one hand we have our economy—the market economy—which knows the price of everything but the value of nothing. That is not its job. And then the state or government which attempts to use its means of acquiring tax money, of making decisions to assist us in becoming a better, more equitable society as it defines it. That is what all societies are currently caught between—forces that are more complex and bigger than any of us can understand. And missing in that equation, as we have political and ideological struggles between those who think market economies are the answer to everything, those who think government programs are the answer to everything, is the recognition among all of us that neither of those is an adequate explanation for the challenges confronting us.

And what we each must do is break through the old thinking that has for too long captured us politically and institutionally, so that we can begin to devise new ways of thinking about not only what it means to have economies that doesn't discard people like they were excess baggage that we no longer need, but to define our institutional and personal responsibilities in ways that answer this lack of meaning.

This, then, brought Mrs. Clinton to the high point of her speech, to her solution, to those key words that would live well beyond that lecture hall. She declared:

We need a new politics of meaning. We need a new ethos of individual responsibility and caring. We need a new definition of civil society which answers the unanswerable questions posed by both the market forces and the governmental ones, as to how we can have a society that fills us up again and makes us feel that we are part of something bigger than ourselves.

She was short on specifics as to precisely what that meant, a lack of elucidation that brought her much criticism. This, too, was unfair, since her point was to introduce a concept, leaving the details, particularly policy specifics, to others, and perhaps even to herself, later. That said, she rather clumsily managed to tie the theme to health care:

But to give you just one example about how this ties in with what I have said before about how these problems we are confronting now in many ways are the result of our progress as we have moved toward being modern men and women: Our ancestors did not have to think about many of the issues we are now confronted with. When does life start; when does life end? Who makes those decisions? How do we dare impinge upon these areas of such delicate, difficult questions? And, yet, every day in hospitals and homes and hospices all over this country, people are struggling with those very profound issues.

These are not issues that we have guidebooks about. They are issues that we have to summon up what we believe is morally and ethically and spiritually correct and do the best we can with God's guidance. How do we create a system that gets rid of the micro-management, the regulation and the bureaucracy, and substitutes instead human caring, concern and love? And that is our real challenge in redesigning a health care system.

She talked about how "discussions" needed to be had, at home, in the workplace, in schools; about taking a "hard look" at the nation's "institutions," media, "values," "challenges"; about "breaking new ground," about even, at one point, the "God-given potential" of every child. Here her speech became very vague, as she drifted from one generality to another, almost to the point that a lot of what she said was understandably perceived as much touchy-feely ado about nothing. The speech probably struck a chord with certain liberals unable to find salvation through politics in the Reagan era, but left far more scratching their heads.

On the other hand, some of the language in the speech was clear enough to be alarmingly similar to other words from only a year earlier—Vice President Al Gore's 1992 environmental manifesto, *Earth in the Balance.* In his book, the man who shared Bill Clinton's presidential ticket urged that the rescue of the environment become the "central organizing principle" of all modern civilization, and that such a struggle would require "a wrenching transformation of society."[18] Like Gore, Mrs. Clinton served up some questionable talk about transforming human nature. In one Gore-like moment, Mrs. Clinton stated in the speech: "Let us be willing to remold society by refining what it means to be a human being in the 20th century, moving into a new millennium."

This was a halting statement. Most Americans preferred not to remold society and redefine humanity; that was a bit beyond what they had voted for in November 1992. Nonetheless, in the final line of her address, she concluded by reiterating her objective of "remolding a society that we are proud to be a part of. Thank you all very much."[19]

Don Jones got a copy of the text and read it over. He right away saw the parallels to Tillich's teachings on alienation and meaninglessness. "These were precisely the terms Hillary struck in that speech in Austin," said Jones. He added surprisingly, "My sense of Hillary is that she realizes . . . that you cannot depend on the basic nature of man to be good and you cannot depend entirely on moral suasion to make it good. You have to use power. And there is nothing wrong with wielding power in the pursuit of politics that will add to the human good. I think Hillary knows this. She is very much the sort of Christian who understands that the use of power to achieve social good is legitimate."[20]

But although Hillary had created a vague manifesto about redesigning the bonds of our society, how her ideology would be received was another matter. This was the first time that she had put forth a broad idea that included both her religious and political ambitions, and the response to it would be a crucial barometer of her future initiatives. While Jones was supportive, that was not the case for most observers of the speech, as the first lady and her staff soon learned as they awaited the political feedback.

Hugh Passes Away

One cannot separate the speech's search for meaning from what was going on in Hillary's personal life at the moment: The two and a half weeks she spent in Little Rock as her father lay on his deathbed led her to give the talk.[21] The very next day after the speech, on April 7, 1993, her father died, as she was returning to Washington from Austin. Sadly, Hugh was unable to give his daughter a final "I love you."[22]

Hugh Rodham was eighty-two years old. His son-in-law, the president of the United States, delivered the eulogy at the funeral. Bill Clinton said that Hugh, a Republican until literally the very end, never gave up hope that his daughter's husband would join him in the GOP and support a cut in the capital gains tax.[23]

Now, Hillary recalled her father's old-fashioned but valuable ethical code of absolutes. "My father was no great talker and not very articulate, and wouldn't have known Niebuhr from Bonhoeffer from Havel from Jefferson," she said, acknowledging that no-nonsense Hugh would have thought a conversation on something like the politics of meaning to be "just goofy"—surely, liberal gobbledygook. "But," she said, "he gave me the basic tools, and it wasn't fancy philosophical stuff."[24]

She harked back: "He used to say all the time, 'I always love you but I won't always like what you do.' And, you know, as a child I would come up with 900 hypotheses. It would always end with something like, 'Well, you mean, if I murdered somebody and was in jail and you came to see me, you would still love me?' And he would say: 'Absolutely! I will always love you, but I would be deeply disappointed and I would not like what you did because it would have been wrong.'"[25]

Now, after all of these years, Mrs. Clinton candidly credited her father for giving her values and the gift of "unconditional love that I think every child deserves to have—and one of our problems is that too many of our children don't have that."[26]

Hugh had given her several foundations. Indeed, some of his values were things that had pushed Hillary to where she was today. Though their politics had hardly been eye-to-eye, he continued to love his daughter, continued to believe in the system of values that he had instilled in her. As she took on new duties unlike any she had experienced before, it was becoming increasingly clear that the Methodism which Hugh was always so passionate about would remain part of his daughter's life. Despite the discrepancies in their practices and party, despite the fact that Hillary's strain of belief would always be more about social salvation than Hugh's rugged individualism, thanks to him she still felt a passion for God.

Amalgam Politics

While Hillary was coping with more than the death of her father, new rumors of her husband's infidelities began to pop up—this time recent infidelities in no less than the Oval Office, where the Clintons had assured everyone the bad behavior would stop for good. As Hillary was out of town, Bill hosted singer-activist Barbra Streisand, who had gone to the White House to give Clinton a private demo of a cut from her upcoming album. Here again, the details are terribly sloppy. Nonetheless, we know that Streisand spent the night in the Lincoln Bedroom. Gail Sheehy suggests that she did not sleep alone. Soon after Hillary's return, says Sheehy, Clinton emerged from his morning jog with a claw mark along his jawline. Press Secretary Dee Dee Myers later sheepishly admitted to Sheehy that she was the "idiot" who covered for the caddish Clinton by saying he had cut himself shaving—before she saw the mark on his face. "Then I saw him," said Myers. "It was a big scratch, and clearly not a shaving cut. Barbra Streisand was clearly around at the time."[27]

Reports like this—founded or unfounded—were not what Hillary needed as she suffered through the demise of her father. In response to the rumors about Bill and Streisand, says Sheehy, Hillary barred Streisand from further overnight appearances at the White House.[28] Beyond Hillary, the rumors were extremely damaging, regardless of their legitimacy, and a great setback to women on the left who viewed Hillary as a political heroine, the first *real* first lady since Eleanor Roosevelt. It was four months into the presidency, and the Clintons were already a soap opera. Hillary, however, battled for Bill's respectability and her own intellectual recognition.

To that end, the first lady's interest in the politics of meaning became a subject of intense curiosity in the press, forcing her to quickly defend and even try to define herself. "My politics are a real mixture," she explained. "An amalgam. And I get so amused when these people try to characterize me: 'She is *this*, therefore she believes the following 25 things.' . . . Nobody's ever stopped to ask me or try to figure out the new sense of politics that Bill and a lot of us are trying to create. The labels are irrelevant."

Hillary insisted that she could not be placed in a box politically or ideologically: "And yet, the political system and the reporting of it keep trying to force us back into the boxes because the boxes are so much easier to talk about. You don't have to think. You can just fall back on the old, discredited Republican versus Democrat, liberal versus conservative mindsets."

To a degree her plea was valid, but it was also disingenuous, as she could have candidly admitted that she was not a conservative Republican. Indeed, when she described herself in a *Parade* magazine interview that April as "conservative in the true sense of that word—not in the kind of radical, ideological, destructive way that term is often used," *Newsweek*'s writer Eleanor Clift rightly criticized her for trying to disguise her true self.[29] This trend would continue throughout Hillary's career, as she consistently lunged for the rhetorical middle and tried to frame herself as a moderate.

Nonetheless, as Martha Sherrill noted in the *Washington Post*, Mrs. Clinton thought herself "a citizen" who spoke of "virtue" and "personal responsibility" and "being connected to a higher purpose." The first lady even offered a surprising nod to the religious right: "Much of the energy animating the responsible fundamentalist right," she said, "has come from their sense of life getting away from us—of meaning being lost and people being turned into kind of amoral decision-makers because there weren't overriding values that they related to. And I have a lot of sympathy with that." She added: "The search for meaning should cut across all kinds of religious and ideological boundaries. That's what we should be struggling with—not whether you have a corner on God."[30]

To her credit, Mrs. Clinton was at least attempting to elevate the public discourse when it came to religion, trying to move away from the one-size-fits-all generalizations about the religious right by many Democrats. Thus, she and her staff were shocked when, on May 23, 1993, the *New York Times Magazine* ran an influential analysis of her and her ideas by Michael Kelly, mockingly titled, "Saint Hillary," with those words irreverently scratched next to a photo of Hillary in a white dress, for which she had posed, to the initial delight of her staff, which was suddenly no longer pleased. She had talked openly and honestly of her values and her faith, and now this headline?[31]

Kelly began by stating that since she was a teenager, Mrs. Clinton had long aimed to help, to see that people live by the Golden Rule. Now her goal was to achieve these things on a larger scale, to "make the world a better place—as she defines better." Kelly noted that although most first ladies have a general warmth and compassion, there was a major difference in the case of this First Lady: She was serious and she had power.

Kelly correctly added that Hillary's sense of purpose stemmed from a worldview rooted in the Don Jones Christianity of her youth and grew from the belief that the baby boomer generation was meant to inform the world of its flaws. These two halves of the coin formed the "true politics of her heart." Yet that was simply the motivation. Where was all this leading Mrs. Clinton? Kelly answered:

> Driven by the increasingly common view that something is terribly awry with modern life, Mrs. Clinton is searching for not merely a programmatic answer but for The Answer. Something in the Meaning of It All line, something that would inform everything from her imminent and all-encompassing health care proposal to ways in which the state might encourage parents not to let their children wander all hours of the night in shopping malls. When it is suggested that she sounds as though she's trying to come up with a sort of unified-field theory of life, she says, excitedly, "That's right, that's exactly right!"

Kelly reported that the first lady was seeking a way to marry conservatism and liberalism, capitalism and state-ism, to join together the myriad state, religious, social, economic, and class problems— from the end of communism to unwed mothers to aggressive panhandlers—into one idea that could be addressed by her theory. He quoted her: "It's not going to be easy. But we can't get scared away from it because it is an overwhelming task." And that difficulty, Kelly assessed, was bound to be doubled by the apparent reality that a good deal of what Mrs. Clinton perceived as wrong with the American way of life could be traced to the liberal baby boomers' failed social experiments to redefine society on their terms.

Kelly dubbed this "the Crusade of Hillary Rodham Clinton," which he said had commenced on April 6 in Austin, Texas. The piece conceded that it was "easy to mock this sort of thing." Yet it did not help, said Kelly, that in their interview Mrs. Clinton continued to "grope" for a better articulation of her thinking. "I don't know; I don't know," she admitted to Kelly when asked to define her philosophy. "I don't have a coherent explanation. I hope one day to be able to stop long enough actually to try and to write down what I do mean, because it's important to me that I try to do that, because I have floated around the edges of this and talked about it for many, many years with a lot of people, but I've never regularly kept a journal or really tried to get myself organized enough to do it."

As she continued to wax philosophical with Kelly, she became more concrete, though Kelly ironically took a few extra shots once she finally seemed to find some bearings. "The very core of what I believe is this concept of individual worth, which I think flows from all of us being creatures of God and being imbued with a spirit," she said. "Some years ago, I gave a series of talks about the underlying principles of Methodism. I talked a lot about how timeless a lot of scriptural lessons were because they tied in with what we now know about human beings." Among those was the Golden Rule and Christ's greatest of all commandments: "If you break down the Golden Rule or if you take Christ's commandment—love your neighbor as yourself—there is an underlying assumption that you will value yourself," added the first lady, "that you will be a responsible being who will live by certain behaviors that enable you to have self-respect, because, then, out of that self-respect comes the capacity for you to respect and care for other people."

At this point, Hillary offered to Kelly some specific examples of how this philosophy could and should play out in everyday life:

And how do we just break this whole enterprise down in small enough pieces? Well, somebody says to themselves: "You know, I'm not going to tell that racist, sexist joke. I don't want to objectify another human being. Why do I want to do that? What do I get out of that kind of action? Maybe I should try to restrain myself."

Or somebody else says: "You know, I'm going to start thanking the woman who cleans the restroom in the building that I work in. You know, maybe that sounds kind of stupid, but on the other hand I want to start seeing her as a human being."

And then maybe the next step is I say to myself: "How much are we paying this woman who works the 3 to 11 shift. And who's taking care of her kids while she's here working? And how do we make it possible for her to be able to both be a good parent and perform a necessary function?"

Kelly judged this passage "rambling" and said that it seemed to validate *The New Republic*'s demand of Hillary's politics: "What *does* it all mean?" Kelly was frustrated that this politics of meaning was "hard to discern under the gauzy and gushy wrappings of New Age jargon that blanket it." Then he was more frustrated, as he asked the first lady if there was one unifying idea at the heart of the politics of meaning, only to have her answer: "I don't think there is one core thing. I think this has to be thought through on a variety of planes. I don't think there is one unifying theory."

Overall, Kelly's profile suggested that she was not entirely sure exactly what she meant. On the other hand, Kelly rightly grasped that below the New Age veneer of Mrs. Clinton's musings was an old and fundamentally American dedication to values. He also discerned that this Hillary Rodham Clinton was a big surprise to the electorate, who had seen her as the radical feminist lawyer, the breadwinning wife of Governor Clinton, and not the moralist who had promoted Methodist teaching from Arkansas pulpits in the 1980s.

And the left and the right were surprised at Kelly's inclusion of her endorsement of calling evil "evil." She told Kelly about a moment when she was sitting in a law school in the early 1970s observing a hypothetical discussion about the motives of terrorists. "And I remember sitting there listening to the conversation as so many people tried to explain away or rationalize their behavior," said Mrs. Clinton. "And I remember saying, 'You know, there is another alternative. And the other alternative is that they are evil. I mean, you know? There *are* evil people in the world. And they may be able to come up with elaborate rationalizations to attempt to explain their evil, and they may even have some reasonable basis for saying their conduct needs to be understood in the light of pre-existing conditions, but their behavior is still *evil*." [32]

The impact of the piece was substantial as the perceived formlessness of Hillary's ideas put her squarely in the crosshairs of public scrutiny. Here Hillary's politics of meaning had been laid bare for everyone on the right and the left, exposing them to dissection and leaving them wide open to criticism from both sides of the aisle. While Republicans might be seeing a value system in the first lady that they did not know about on Election Day, Democrats were seeing the same thing—and the results on both sides were not necessarily a good thing.

Though Hillary with her newfound visibility was, in some sense, wearing her faith on her sleeve, the question of political cost remained. She had created a hybrid belief system, incorporating ideas from both the right and the left, but without enough of either to gain true supporters or make her an actual moderate. The end result was that the more people seemed to uncover about the intersection of her faith and politics that existed in her politics of meaning, the less they seemed to like it. Unfortunately for Hillary, this was a trend that would only increase.

Michael Lerner Speaks Up

While there was certainly an inspiration of Methodist social gospel in Hillary's controversial new political-spiritual thinking, there was more the influence of something else—actually, *someone* else. Where had this politics of meaning come from? *Who* had it come from?

The source of these ideas was a man named Michael Lerner, who on June 8 flew into Washington to explain himself, his ideas, and his and Hillary's meaning. He needed to do so, since over the last two months he had been called Hillary Rodham Clinton's "guru" (and much more) amid the ensuing "press riot," as he called it. Even normally sympathetic sources on the left were puzzled. *The New Republic* had asked, "What on earth are these people talking about?" *Time* magazine seconded, "The politics of what?" And a *Los Angeles Times* piece employed the word "psychobabble."[33]

Lerner was the editor and publisher of the progressive *Tikkun* magazine, what Henry Allen of the *Washington Post* described as "a leftish bimonthly published out of the fifth floor of a synagogue on New York's Upper West Side," and whom Allen described as "a scruffy hypomanic left-wing New York intellectual."[34] Lerner grew up in Newark, went to a private school, was the son of a lawyer and a mother who worked for Senator Harrison Williams. John F. Kennedy was a family friend, so much so that he wrote Lerner's college recommendation for Columbia.

By 1964 he made his way to Berkeley, where he became part of the free speech movement and Students for a Democratic Society. He got a Ph.D. in psychology at Berkeley and went on in 1977 to help start the Institute for Labor and Mental Health, where he focused on the "psychodynamics of the working class."[35] Lerner was a self-described psychotherapist and researcher studying "working-class conscious-ness."[36] More than one biographer noted that at his wedding to his first wife, in a reportedly nonlegal ceremony, Lerner cut into a cake with the inscription "Smash Monogamy," after the couple exchanged rings hammered out of metal from downed U.S. military aircraft.[37]

With this résumé, Lerner was a peculiar choice of a spiritual-political mentor to help Hillary shape public policy. In a time when both she and her husband were trying to regain the center and con-vince America that a New Democrat could lead this country more effectively and moderately than a Republican, Lerner was a curious throwback to the Clintons of the 1970s. He was progressive, but in an eccentric way that seemed to cast his ideas far on the left, while leaving both him and the first lady vulnerable to caricature.

On June 8, 1993, he held an invitation-only press conference at a Jewish student center at George Washington University. About six reporters accepted the invitation.[38] At the conference, Lerner talked openly about his special relationship with Hillary. After the Uni-versity of Texas speech, he saw Hillary at a White House reception on April 21, 1993, the day before the dedication of the Holocaust Memorial Museum. Hillary looked directly at Lerner and said with a big grin, "Am I your mouthpiece or what?" Lerner told her that he had specific ideas about how to turn the politics of meaning into actual policy. According to Lerner, the first lady responded, "Great, let's talk about it. Well, can you come back soon?"[39]

The following Monday, said Lerner, they were meeting in the first lady's office, where they spent about half an hour together. Said Lerner: "The conversation was quite amazing . . . almost like in half-sentences because she had read everything. I mean she had under-stood everything."[40]

Lerner informed the reporters that this was hardly their first con-tact: Bill Clinton had been reading Lerner since 1988, when he sent Lerner a letter from the Arkansas governor's mansion, telling him he had "helped me clarify my own thinking." Hillary confirmed to

Lerner that "Bill is just as interested in the politics of meaning as I am."[41] Later, in 1993, once Bill became president, Lerner went directly to the new president with his policy ideas. In a memo to President Clinton, according to Henry Allen, Lerner sweepingly proposed that every government office be required to justify budget requests by answering questions like: "How do our programs foster caring, concern for others, ecological awareness, spiritual sensitivity, and a sense of mutual responsibility?" For example, "the Department of Labor should create a program to train a corps of union personnel, worker representatives and psychotherapists in the relevant skills to assist developing a new spirit of cooperation, mutual caring and dedication to work." In the memo, he proposed a "summit conference on ethics, community, and the politics of meaning."[42]

Here were some specifics, some flesh to the bones of the politics of meaning. On June 13, 1993, Lerner went further in an op-ed piece in the Outlook section of the Sunday *Washington Post*, in which he explained:

> The point of the politics of meaning is not that government should dictate a particular moral or spiritual view. But we live in a time when our economy rewards the self-centered and the selfish while putting at a disadvantage those who have taken time off in order to act morally or in a caring way toward others.
>
> A progressive politics of meaning seeks to level the playing field by creating economic, political and social incentives for social and ecological responsibility—so that people can feel that choosing a moral life does not mean losing all chance for personal advancement, economic reward or social sanction.

Lerner then explained that the politics of meaning was his, and by implication Hillary's vision for the 1990s, an idea that might come to define the whole of the Clinton administration. Here he seemed to be venturing toward the same territory that Hillary had traveled on when she spoke to Michael Kelly. Lerner articulated the politics of meaning as something that transcended all problems and struck at the heart of all Americans' struggles. It was about the activities that people engage in—drugs, TV, exercise—to distract them from the problems in their lives. It was about how these solutions were merely approximations. It was about the weakening of connections between people and in relationships. And it was about how the marketplace was laughing all the way to the bank.

From these generalities, Lerner went on to highlight specific policy goals that could be implemented; whereas Hillary had avoided tying her politics of meaning to individual proposals, Lerner dived into these headfirst. He suggested offering financial incentives to encourage companies to be more socially and ecologically conscious. He recommended the creation of worker health committees to ensure that the stress levels of workers remained in check. On the education side, he proposed that schools put as much emphasis on teaching empathy as they do on teaching math.

While Lerner's ideas might have been far-fetched, they were beginning to bring the politics of meaning into greater focus, and many people did not like what they saw.

The bigger picture was not simply a "hunger for larger purpose," but a sudden fear by many that the political left, which had failed to mold society through the public sector, was now seeking a giant step beyond the New Deal and Great Society; it looked like the left might now be seeking to use the power of the federal government to force the private sector into achieving the left's vision. This was a legitimate interpretation of Lerner's details.

Lerner did not acknowledge such a goal, or state it that way, and neither did the first lady, but that was a practical effect. The politics of meaning could place all of society behind a blueprint for a desired social model, one that would presumably give meaning to the lives of the "society shapers" on the left who felt morally impoverished.

In short, the politics of meaning was a signal that while the religious right was content with churches and faith-based organizations changing the culture and society, the religious left—or at least Lerner and Mrs. Clinton—seemed to want a spiritually motivated "caring" government to do the job at every level. Though neither Lerner nor Hillary addressed how such an expansion would take place, there was no doubt that the sweeping reform they discussed would usher in an era of federal power potentially more grandiose than even the New Deal. This was the kind of thing that most Americans—distrustful of overly expansive government—were not likely to embrace. And by the end of the midterm elections in 1994, the public would make this point all too painfully clear for the Clinton administration.

CHAPTER 8

The Clintons, the Pope, and Mother Teresa

Perhaps the most interesting political-religious relationships for Hillary and Bill Clinton in the 1990s came not with Michael Lerner but with the world's two most prominent Catholics—Pope John Paul II and Mother Teresa. What made these relationships interesting—at times almost contentious—was the subject of abortion.

As many of the top pro-life Protestants were kept out of the Clintons' orbit, the pope and Mother Teresa were the two highest-level pro-life ambassadors with whom the Clintons met in the 1990s. The very positions and profile of these two leading Catholics meant that the Clintons could not and did not refuse to meet with them, nor could the Clintons refuse to at least listen to the messages on abortion that they insisted on delivering, directly and with no apologies.

Meeting John Paul II

The first of their encounters with John Paul II came in August 1993. From August 12 to 15, 1993, the pope came to America for the fourth international World Youth Day, held in Denver, Colorado—his first trip to America during the young presidency of Bill Clinton. On Thursday, August 12, the pontiff stepped off the plane at Denver's Stapleton International Airport, ready to speak to a throng of faithful. He was greeted by President Clinton, the first lady, and the first daughter. The president offered welcoming remarks.

According to his biographer, the pope departed from his prepared text after hearing what Clinton had to say, and did so in a pointed way.[1] He first thanked the president, Mrs. Clinton, and Chelsea for their "kind gesture in coming here personally to welcome me," and then expressed gratitude to the young people who were present. The pope said that this would be a World Youth Day for "serious reflection on the theme of life: the human life, which is God's marvelous gift to each one of us."

The pontiff then got to the essence of what he felt was the critical message that America and its first family needed to hear: He said that America had been founded on the assertion of certain self-evident truths concerning the human person, including the inalienable right to life of every human being. Now, with the Cold War over, said the pope, all the "great causes" led by the United States "will have meaning only to the extent that you guarantee the right to life and protect the human person." He said that the "ultimate test" of America's greatness was the way that Americans treated "every human being, but especially the weakest and most defenseless ones. The best traditions of your land presume respect for those who cannot defend themselves." He then raised his voice: "If you want equal justice for all, and true freedom and lasting peace, then, America, defend life!" There was no doubt by anyone in the crowd that the words were aimed not just at the nation but at its president and first lady standing at the pontiff's side.[2]

As one of the major forces in the fall of communism, John Paul II was announcing that his role in that great drama would not be his only legacy. The civilized world still possessed other battles that needed fighting, other work that needed to be done. The answer came in the mid-1990s with the publication of his encyclical *Evangelium Vitae*, the *Gospel of Life*.[3] The Western cultures that had legalized abortion were now speeding toward euthanasia—killing unwanted, "inconvenient" human beings at the other end of life—and embryonic research, destruction at an even earlier stage of human development. Pope John Paul II coined a phrase that he believed summed up the crisis: The next great moral battle, the next crisis spearheaded by the forces of secular atheism, was the struggle for life; the world was confronted with a choice between a "Culture of Life" and a "Culture of Death."

For his part, Bill Clinton, in August 1993, surely understood that he was presiding through a major transitional era.[4] Given their pro-choice views, there is no evidence or reason to suspect that he or Hillary shared the pope's diagnosis. Notably, however, the pope spoke in language that must have struck Hillary, pushing all her social justice buttons. To protect human life, the pope said, was for America to serve its noblest ideals, its destiny as one nation, under God, "that protects and extends liberty and justice to all."[5] These were tough words for Hillary to swallow, as they so closely mirrored her belief in justice for all classes and ethnicities that she extrapolated from her own faith. In this case, however, the pope opened the social justice umbrella to encompass unborn human life, an expansion that she was unwilling to replicate.

Asked afterward if he and the pontiff had much discussion, Bill Clinton said that they had discussed "social issues." Neither he nor the first lady cared to elaborate. This would be the first of many times that faith and values would collide for the Clinton administration and the Vatican—always they would split over these life issues; always the pope would remain steadfast.

The First Lady and the Lady from Calcutta

A few months later came another prominent Catholic to bring a similar message that the Clintons did not want to hear: Mother Teresa. Nevertheless, this was a Catholic with whom Mrs. Clinton would develop a somewhat close relationship, even a working relationship in one case.

A key development had occurred on the abortion front in the intervening months between John Paul II's trip to America and Mother Teresa's scheduled visit six months later: Mrs. Clinton, by then full throttle in her efforts to revolutionize the American health care industry, said in an October 1993 televised forum that, under her plan, abortion services "would be widely available." This prompted anxieties over the prospect of taxpayer-funded abortions, sparking the Coates Amendment in the U.S. House of Representatives, which sought to strip abortion funding from the plan. Mrs. Clinton's intentions sent pro-life Democrats like Pennsylvania governor Robert P. Casey into such a rage that Casey considered a run for the presidency to dislodge the Clintons.[6]

In her remarks, the first lady allowed for a "conscience exemption" in which doctors and hospitals would not be forced to perform abortions.[7] Pro-lifers were relieved on that; still, they could not fathom that their tax dollars might be used to fund what they saw as the deliberate destruction of innocent human life. More, they feared the "abortion clinic mandate" in Mrs. Clinton's plan, which would suddenly require the availability of abortion services in many of the 87 percent of United States counties that did not have clinics at the time.

Mrs. Clinton's words in October also ignited fears among moderate and conservative Christians over the availability of the abortion pill, RU-486, under her health care plan. One of her husband's first acts in office was to push the pill to market through an expedited FDA approval process that was criticized by pro-lifers as allegedly too quick for the safety of the women who would take the pill.

This had been a longtime goal of abortion advocates like Ron Weddington, who with his wife, Sarah Weddington, had presented "Jane Roe" to the Supreme Court. Ron Weddington sent a four-page letter to President Clinton, urging: "I don't think you are going to go very far in reforming the country until we have a better educated, healthier, wealthier population." The new president could "start immediately to eliminate the barely educated, unhealthy and poor segment of our country." "No, I'm not advocating some sort of mass extinction of these unfortunate people," wrote Weddington. "The problem is that their numbers are not only not replaced but increased by the birth of millions of babies to people who can't afford to have babies. There, I've said it. It's what we all know is true, but we only whisper it." By "we," Weddington said he meant "liberals," like himself and the Clintons.[8]

There is great irony in Weddington directing these sentiments to the Clintons: Consider, after all, that those "unfortunate people" who could not afford to have babies once included Hillary's maternal grandmother and Bill's own widowed mother, both of whom, by choosing life under financially difficult circumstances, allowed Hillary Rodham and Bill Clinton to enter this world. Weddington pointed to the Clintons themselves as the "perfect example" of how to reproduce responsibly, unlike the baby-birthing "religious right": "Could either of you have gone to law school and achieved anything close to what you have if you had three or four or more children before you were 20?" he asked, urging passage of RU-486. "No! You waited until you were established and in your 30's to have one child. That is what sensible people do."[9]

Bill Clinton did something else that Weddington considered sensible: He authorized RU-486. Mrs. Clinton was thrilled with the move.

To Mother Teresa, these were tragic victories for what her pope would call the Culture of Death. In February 1994, they stoked her worst fears about where American health care policy was headed under the Clintons, as she prepared to meet the partners. The occasion was the annual National Prayer Breakfast, a popular gathering of ecumenical worshippers and officials from Washington. As president, Bill Clinton attended every one of them, with Hillary usually (if not always) accompanying him.[10] This particular year, on February 3, 1994, the keynoter was Mother Teresa, a Nobel winner and saintly figure who had come all the way from the most impoverished area of the planet, the slums of Calcutta. According to Kathryn Spink's later authorized biography, the reluctant nun was invited and encouraged by President Clinton himself.[11]

Held at the Hilton, the breakfast was attended by nearly three thousand people packed into the huge room—Jews, Muslims, Buddhists, Hindus, Catholics, all forms of Protestants, even agnostics and atheists, and the press, including C-SPAN's cameras. Near the dais were the president and first lady, along with the vice president and his wife, and a select few VIPs, including Supreme Court justices and the highest ranking members of Congress. Unlike in typical years, when the keynoter sits among the assembled and waits for others to finish before his or her turn, Mother Teresa emerged from a curtain behind the platform only when she was called, and then slowly hunched her way to the microphone. Hillary was struck by how tiny she was, wearing only socks and sandals in the bitter cold.[12]

The title of her talk was, "Whatever You Did Unto One of the Least, You Did Unto Me." She began by talking about Jesus and John the Baptist in utero, about their mothers, Mary and Elizabeth, and how the "unborn child" in the womb of Elizabeth—John the

Baptist—leaped with joy as he felt the presence of Christ in the room when Mary entered to speak to Elizabeth.

Hillary might have seen what was coming, since she herself might well have employed the same imagery at a United Methodist venue, the same story, but it would have been to make a point about child care or even discrimination or economic fairness. That was not the direction where Mother Teresa was leading.

She next spoke of love, of selfishness, of a lack of love for the unborn—and a lack of want of the unborn because of one's selfishness. Jesus, who, said the nun, had brought joy while still in the womb of Mary, had died on the Cross "because that is what it took for Him to do good to us—to save us from our selfishness in sin."[13]

Mrs. Clinton could relate to this; in Michael Lerner's *Washington Post* op-ed laying out his politics of meaning, he had explained that both he and Mrs. Clinton insisted that "society can and ought to move from an ethos of selfishness to an ethos of caring and community."[14] So, yes, Hillary understood Mother Teresa's point. Again, though, she would have applied the parable to something like race or class, not abortion.

Peggy Noonan, the former Reagan speechwriter and a pro-life Catholic, was there. She says that by this point in the talk, the attendees began shifting in their seats, as a lot of what the lady from Calcutta had to say was striking too close to home. Then the nun said something that made everyone very uncomfortable: "But I feel that the greatest destroyer of peace today is abortion, because Jesus said, 'If you receive a little child, you receive me.' So every abortion is the denial of receiving Jesus, the neglect of receiving Jesus."[15]

Here, Noonan described a "cool deep silence" that enveloped the room, but only for a very brief moment, and then applause started on the right side of the room, and then spread throughout the crowd, as people began clapping and standing; the ballroom was swept up in a nonstop applause that Noonan says lasted for five to six minutes.[16]

Yet some did not clap at all. Hillary Clinton did not, and neither did her husband, nor Vice President Al Gore and Tipper Gore. They sat there, in the glare of the hot lights, all eyes in the crowd fixed upon them as they tried not to move or be noticed, conspicuous in their lack of response, clearly uncomfortable as the applause raged on.

Their lack of approval was puzzling. Though pro-choice, the Clintons and the Gores had said many times that abortion is regrettable, a terrible choice, and even wrong, and should be limited. So on some level, they must have agreed with what Mother Teresa said, or at least one would think so. She condemned abortion, said Jesus would abhor it, but she did not call for repealing abortion laws, even though she would support such action.

The tiny, weak, aged lady was only warming up. She had seen and experienced real suffering and could not care less about making momentarily uncomfortable a crowd of a few thousand financially comfortable people who had never known real material deprivation, and whose only crisis each morning was traffic or a line at Starbucks. She returned to that selfishness point: "By abortion, the mother does not learn to love, but kills even her own child to solve her problems." Abortion was "really a war against the child, and I hate the killing of the innocent child, murder by the mother herself. And if we accept that the mother can kill even her own child, how can we tell other people not to kill one another? . . . Any country that accepts abortion is not teaching its people to love one another but to use violence to get what they want. This is why the greatest destroyer of love and peace is abortion."

The Aftermath of Mother Teresa

Throughout the talk's high points on abortion—the raw nerve—the Clintons and Gores remained in stony silence. Said one attendee at the event, a pro-life Catholic and high-level appointee in the Reagan Administration, who asked not to be named: "It was an outrage, an abomination, very rude. Mrs. Clinton in particular just sat there. I will never forget that moment. It told me all I needed to know about her."

Hillary Rodham Clinton took her lumps for this one. So did her husband. And to his credit, Bill realized that the behavior of him, his wife, and the others was rude. According to Kathryn Spink, he apologized to Mother Teresa after the speech.[17] Hillary responded later that day—sort of. In commenting on Mother Teresa's remarks, she must have briefly given the nun hope that she, too, would speak on behalf of the unborn when she began, "I have always believed that Christ wanted us to be joyous, to look at the face of Creation and to know that there was more joy than any of us could imagine." As the Champion of Calcutta held her breath, however, she was disappointed, as Mrs. Clinton did what she has long done; she applied the thought very selectively, restricted it solely to economics, not unborn life, as she followed: "Or as Mother Teresa told us this morning, to see the joy on the face of a homeless beggar, who is picked up off the street and brought in to die, says joyously, 'Thank you.'"[18]

Hillary's statement was an extraordinary example of the compartmentalization mastered by both her and her husband, in which they ignored the obvious and justified a cause through spiritual means that failed to take into account the clear, intended purpose. The reaction was surreal. Mother Teresa had come to give a major moral statement on abortion, and did so in a way that shocked the entire crowd. And then Mrs. Clinton literally ignored the entire message in her follow-up remarks, carefully lifting a smaller item from the nun's address, one with which she agreed; then placed it fully out of its context, and used it for an entirely separate political purpose with which Mrs. Clinton was politically satisfied.

And it was not as though Mrs. Clinton did not get the point.

"She [Mother Teresa] had just delivered a speech against abortion," explained Mrs. Clinton in assessing the keynote address almost ten years later. In the minutes after the talk, said Hillary, the nun persisted, taking the abortion issue directly to Hillary's face: "[S]he wanted to talk to me," said the first lady. "Mother Teresa was unerringly direct. She disagreed with my views on a woman's right to choose and told me so."[19] In other words, there was no mistaking the message that day, or that Hillary got it unerringly.

On the other hand, Hillary later, perhaps upon further reflection with the help of an aide, identified a crucial component of the speech that she did not need to take out of context to find common ground. Mother Teresa had said: "Please don't kill the child. I want the child. Give me the child. I'm willing to accept any child who would be aborted and to give that child to a married couple who will love the child and be loved by the child." Echoing the Malcolm Muggeridge phrase that introduced her to the West, Mother Teresa said, "I will tell you something beautiful. We are fighting abortion by adoption."

Now that was something that Hillary could applaud. In the course of one of their subsequent conversations, Mrs. Clinton made clear that while she supported legalized abortion, she also wanted to see more adoptions as an alternative. The nun told the first lady she had placed more than three thousand orphaned babies into adoptive homes in India. Hillary said she would like to visit the orphanage in New Delhi. Several months later, she and Chelsea did just that, visiting one of the Missionaries of Charity homes in New Delhi, a facility, said a lawyerly Hillary, that "would not have passed inspection in the US" because there were too many cribs crowded together.[20]

Mother Teresa informed the first lady of her goal of establishing a home in Washington, D.C., where mothers could take care of their babies until they found adoptive or foster homes. In turn, Hillary went to bat for her, rounding up pro bono lawyers to do legal work, fighting through the bureaucracy of the District of Columbia, and doing what she could to lend a hand to what became the Mother Teresa Home for Infant Children near Chevy Chase Circle, just over the Washington, D.C., line. She telephoned community leaders and pastors from nearby Baptist, Catholic, Episcopal, and Presbyterian churches, calling them to the White House to see where and how they could help. Moving the bureaucracy, Hillary later said, turned out to be harder than she had imagined.[21]

Mother Teresa was equally relentless on her end. When she felt the project was lagging, she sent a letter to the first lady checking on the progress. "She sent emissaries to spur me on," recalled Hillary. "She called me from Vietnam, she called me from India, always with the same message: When do I get my center for babies?"[22]

On June 19, 1995, the shelter for children opened, the Mother Teresa Home for Infant Children. This led to a photo op of Hillary and Mother Teresa clasping hands in the newly decorated nursery and smiling at each other. A reporter could not resist asking the uncomfortable question: Yes, conceded the first lady, of course they had discussed their "philosophical differences" over abortion. Mother Teresa, ever the peacemaker, stepped in to underscore where the focus should be at that particular moment, namely, on where they agreed: "We want to save the children," she said. The nun, slow and frail, held Hillary's arm as they toured the facility, examining the freshly painted nursery and rows of bassinets awaiting infants.[23]

This was not the end of the relationship, which Hillary looks back upon with fondness. In the short time she had left on earth, Mother Teresa continued to try to change Mrs. Clinton's view on abortion. According to Hillary, "she sent me dozens of notes and messages with the same gentle entreaty." She dealt with the first lady with patience and kindness, but firm conviction: "Mother Teresa never lectured or scolded me; her admonitions were always loving and heartfelt," wrote Hillary, adding that she had "the greatest respect for her opposition to abortion."[24] Mother Teresa saw in Hillary a potentially huge convert to the pro-life cause, and as was her style, she never gave up hope.

Two years after their tour through the foster home in Chevy Chase, on Friday, September 5, 1997, Mother Teresa's frail heart beat its last. The funeral Mass was held at St. Thomas Church in Middleton Row, Calcutta. Hillary was there. After the memorial service, she unexpectedly found herself invited to a private meeting at the mother house, the headquarters of the order founded by Mother Teresa. As the nuns formed a circle around the coffin, where they stood in silent meditation, one of them, Sister Nirmala, Mother Teresa's successor, asked the first lady if she would offer a prayer. Later confessing to feeling inadequate to do so, Hillary hesitated and then bowed her head and thanked God for "the privilege" of having known this "tiny, forceful, saintly woman."[25]

But that final good-bye to the lady from Calcutta would come later. For now, the difficult experiences of 1993 and 1994, including the blistering attacks on Hillary's plan to "socialize" medicine in the United States, were taking their toll on the first lady.

This was seen up-close by her Little Rock minister, Ed Matthews, who paid a visit to Hillary shortly after the politics of meaning issue exploded, when many were satirizing her as a sappy "New Age" believer that had never been a conventional Christian, let alone a serious one. Matthews informed her that some back home at the Methodist church in Arkansas were demanding he expel her from the congregation, as if she was a heretic to the faith of John Wesley. He even claimed to have faced threats on his life.

Hillary had seen worse. She told Reverend Matthews that if he wanted to see hardship, he should sit in her seat. "This is hard," she told the minister. "I'm having a more difficult time . . . than the President."[26]

CHAPTER 9

The Debacle of November 1994

Ideas have consequences. And Hillary Rodham Clinton had embraced Michael Lerner's ideas just as she began working on the task force that was reexamining the way health care was delivered in America. Lerner himself said that his politics of meaning needed to be applied to "the most visionary elements of [the] Clintons' plans for health care."[1] Mrs. Clinton was now ready to do exactly that, in what would become the most far-reaching and controversial element of the first two years of the Clinton presidency, one that would cost the president dearly in the 1994 midterm elections.

The 1994 midterm elections were extremely significant, as the Democrats feared losing both chambers of Congress for the first time in fifty years, foiling everything the Clintons had dreamed about: Without a Democratic Congress, President Clinton's and Hillary's legislative hopes would get nowhere. And what ultimately transpired in 1994 would silence Hillary policy-wise for the remainder of her husband's presidency.

Among the many Democratic candidates in this election year was

Hugh Rodham Jr., who was seeking a Senate seat in Florida. The former Penn State quarterback had his work cut out for him. And his big sister was happy to lend her political weight to the cause. Hillary's little-noticed role in her brother's campaign held important lessons for how she would come to integrate religion into her own later campaigns.

Hugh had given special attention to Florida's Jewish population, which was his single greatest hurdle in securing the Democratic nomination. His Democratic challenger was Mike Wiley, a radio talk-show host who had changed his last name from Schreibman, a point underscored by Hugh Jr. Wiley complained that Rodham was trying to caricature him as an "anti-Semitic Jew." A supporter of Hugh Rodham Jr. had publicly asserted that Wiley had changed his name because he was embarrassed by his Jewish roots—a charge that incensed Wiley.

Into the picture came Hillary, who flew to Florida to campaign for her brother at a synagogue, where they attended a double bar mitzvah. This was the first of many future examples of Hillary openly campaigning for a U.S. Senate seat in a house of worship—an action that the political left usually insists should not be permissible, viewing it as a transgression of the alleged barrier separating church and state. It was fitting that Hillary's campaign work inside the synagogue received almost no major media attention, and certainly no editorials from the *New York Times* crying foul about the first lady's melding of faith and politics. As we shall see, this was the first in a long line of hypocrisy by the mainstream press in its lack of reporting on Mrs. Clinton's open use of religion for political advantage.[2]

Also in 1994, there was big news for Hillary's other brother, Tony, who on May 28 married Nicole Boxer, daughter of Senator Barbara Boxer (D-Calif.), one of the most liberal members of the Senate. They were married in a ceremony at the White House, the first wedding there since that of President Nixon's daughter in 1971. Tony went on to work for the Democratic National Committee, coordinating constituency outreach. All three of Hugh Rodham's children had become liberal Democrats.

Clash with the Vatican in Cairo

As the midterm elections drew near in November, the last thing the Clinton administration needed was another raw issue to anger the moderates who had elected Bill in 1992. But that was precisely what came down the pike in August and September.

Since literally day one of the Clinton presidency, the Vatican had been outraged by the president's actions on abortion. Now, a showdown was scheduled, for September 5–13, 1994. The venue: the World Conference on Population Development in Cairo.

When the conference had gathered ten years earlier in Mexico City, the Reagan administration pushed hard and successfully for a statement unequivocally affirming that legitimate "family planning" did not include abortion. In the intervening years, global activists had searched for an opportunity to turn the tables. Now, with the Clintons in the White House, they had their chance, and a vice president named Al Gore would be the point man, with Hillary offering moral support (and then some) behind the scenes.

The Clinton administration had an ambitious agenda for the September meeting, and was marching in lockstep with international abortion rights groups like Planned Parenthood International. The Vatican was convinced that the U.S. delegation was using slippery, ambiguous language to try to establish an internationally defined and enforceable "human right" to abortion on demand.

As the date of the conference neared, the argument between the two sides got personal, and the Clinton administration looked like it might be backtracking because of the political hit it was taking from moderate to conservative Catholics and other pro-life Christians. The issue came to a head on August 25 when Vice President Gore, speaking to the National Press Club, stated definitively that the United States "has not sought, does not seek, and will not seek to establish an international right to an abortion."

Typically, such an emphatic statement would have resolved the matter. However, this Vatican, and this pope, did not trust this administration on this issue. In an extraordinary counterresponse six days later, the pope's spokesman, Joaquin Navarro-Valls, accused Gore of bad faith, stating: "The draft population document, which has the United States as its principal sponsor, contradicts, in reality, Mr. Gore's statement." This Vatican charge of "misrepresentation" by Gore landed on the front page of the *New York Times* and newspapers around the world.[3] Importantly, the Vatican rarely mentioned a politician by name; singling out "Mr. Gore" was highly uncharacteristic.

The pope's spokesman pointed to specifics, stating that the official Cairo draft document featured a universal definition of "reproductive health care" that included the words "pregnancy termination." This definition, claimed Valls, had been a U.S. initiative. According to the pope's biographer, John Paul II "was not displeased" with this sharp departure from the Vatican's conventional diplomatic reticence.[4] The Vatican was fed up, as was, presumably, the pope.

Mrs. Clinton could not have been happy with the way all of this was unfolding. The White House was taking a political hit, and there was not the victory for women's rights that she had hoped for at Cairo. The cause of international abortion rights would need to wait another day.

Hillary Shares the Faith

There was a further political problem for the first lady: Cairo was another indicator that she was not a moderate. That positioning was crucial to Mrs. Clinton. She knew it had helped her husband win the presidency, and it was vital to her own future political prospects.

Shortly before the November vote, in this increasingly bitter midterm election in which moderates, particularly religious Democrats, were set to make a big difference, Mrs. Clinton consented to a major interview on her faith with *Newsweek*'s religion editor, Kenneth Woodward, published October 31, right before the Tuesday vote.[5]

The piece began by noting that Mrs. Clinton had been called many things, from the right's despised radical feminist to the left's woman-genius. Yet, long before she was a Democrat, a lawyer, or a Clinton, wrote Kenneth Woodward, Hillary Rodham was a Methodist. Woodward noted that she talked like a Methodist, thought like one, and even desired to reform society just like a well-schooled Methodist churchwoman. "I am," she affirmed to *Newsweek*, "an old-fashioned Methodist."

In an otherwise fine article, Woodward may have exaggerated for effect when he claimed that "the Clintons are perhaps the most openly religious first couple this century has seen." It was a faith that Mrs. Clinton, said Woodward, in the wake of the failure of health care reform, and of pundits on her side of the aisle dubbing her politics of meaning to be "flaky New Age blather," had turned to as she felt more and more like "a battered woman."[6] In fact, Hillary was feeling so pummeled in this political season that she claimed to feel a kinship with the God-fearing evangelicals that some of her liberal friends stoned for fun: All of this had taught her, reported Woodward, a "great deal of sympathy" for fundamentalist Christians: "Like them, she believes she has to prove that she isn't a figment of other people's prejudices."[7]

At one point in the interview, the first lady acquiesced to a pop quiz on the nuts and bolts of her Christian faith:

Woodward: "Do you believe in the Father, Son and Holy Spirit?"
    First Lady: "Yes."
    Woodward: "The atoning death of Jesus?"

First Lady: "Yes."
Woodward: "The resurrection of Christ?"
First Lady: "Yes."[8]

Newsweek sought to suggest in the article that Mrs. Clinton was not doing the interview for political reasons as the vote approached. The magazine explained that for three months it had sought an interview concerning her faith, and her aides were fearful that no matter what she said, the first lady would be accused of trying to manipulate her public image for political gain. That clarification, however, may have had the opposite effect, leaving cynical readers to wonder why after three months the first lady and her staff suddenly had a change of heart just as that crucial Tuesday in November finally approached, when the Democrats had reached their lowest point in the polls in a half century.[9]

Mrs. Clinton told Woodward that she kept in her private quarters a copy of *The Book of Resolutions of the United Methodist Church*, along with the Bible. Curiously, she told Woodward: "I think that the Methodist Church, for a period of time, became too socially concerned, too involved in the social gospel, and did not pay enough attention to questions of personal salvation and individual faith." This was an odd comment coming from Hillary, who took Methodism's social gospel more to heart than any other religious teaching. And indeed, Woodward aptly noted that "Mrs. Clinton's many speeches defending the universal healthcare coverage resonated with the moral rhetoric of resolutions adopted by the [Methodist] church's governing General Conference."[10]

Doing what the Clintons do at campaign time, she made overtures to moderates, including on the issue about which she would always be most immoderate, telling *Newsweek* that she believes abortion is "wrong," but, like her husband, added, "I don't think it should be criminalized." But this was not really a concession, as the prolife movement does not favor jailing women who have abortions—

though both Clintons often raise the concern.[11] Pro-lifers would fine and imprison doctors who performed abortions illegally, but the jailing of mothers is not a goal in the pro-life movement.[12]

Another interesting insight into Mrs. Clinton's faith was what she told *Newsweek* about the religious authors and books she was reading at this juncture in her busy life. She had time for only one magazine, the respected weekly *Christianity Today*, and authors such as Father Henri Nouwen, the Reverend Gordon MacDonald, and Tony Campolo.[13] The irony was that the latter two gentlemen would soon be ministering to Mr. Clinton much more so than Mrs. Clinton.

Midterm Defeats

A few days after this interview was published, Hillary's husband and party experienced a stunning defeat, as not a single Republican incumbent in the country—at the level of congressman, senator, or governor—lost an election to a Democrat. In addition, they picked up seats from the Democrats, taking both chambers of Congress for the first time in half a century. In a backlash against the Clintons and the Democratic Party, the GOP accomplished what it could not achieve at the zenith of Ronald Reagan's popularity. The Democratic Party watched some of its best get booted from Congress, including rising stars like Representative Dave McCurdy (D-Okla.) and Representative Stephen Solarz (D-N.Y.). Among those defeated was Hugh Rodham Jr., who had managed to secure the Democratic Party nomination but lost in a landslide to incumbent Republican Senator Connie Mack—70 percent to 30 percent.

There were a number of reasons for the Democrat loss, including the brilliant tactics of House Republican Newt Gingrich (R-Ga.), who crafted a seven-point "Contract with America" that promised the American public a vote on a list of top conservative ideas that all polled at approval ratings of roughly 75 percent, and that he gave maximum exposure by ingeniously publishing it not in the *New York Times* but in *TV Guide*, a widely read publication accessible to most Americans.

But the main reason for the crushing defeat was the leftward lurch of the new Clinton administration, most manifest in the income tax hike earlier in the presidency and, especially, in what was widely perceived as Mrs. Clinton's way left-of-center "nationalization" or "socialization" of the health care industry, which was said to constitute a "government takeover" of one-seventh of the nation's economy.

The morning after, a forlorn President Clinton stepped to the podium to humbly explain that he had gotten "the message." He responded by hiring Dick Morris, the Republican adviser who in Arkansas taught him how to identify a few core middle-of-the-road initiatives, highlight them, and use them to get reelected. This was crucial advice, and it would work—Bill Clinton would not lose again, including in 1996, despite the fact that in November 1994 he looked like a certain one-term president.

Unfortunately for Hillary, this thrust to the center meant "policy death" for her. She was not allowed near health care or any major policy initiative again. She would need to wait six years, until she herself became a candidate.

A stoic Hillary regrouped, looking to the heavens and proudly declaring that she would "measure my choices not against the moment but against eternal values and what is significant for my life. Because that is what I'll be responsible for in the end. . . . It is most important to remain grounded in who you are and what you stand for. That's the only thing that saves you."[14]

Even more unfortunate for Hillary Clinton was that this eventually meant that her name would again be thrust before the public not for anything to do with Bill's policies but instead for Bill's peccadilloes. If she thought the first two years of the presidency had been tough, she had no idea what was in store for her.

CHAPTER 10

New Agers and Eleanor's Ghost

A practical policy result of the Clinton strategy to make moves to the middle was the historic 1995 welfare reform initiative between Bill Clinton and the new Republican Congress, which sought to decentralize the way that welfare was delivered. To this day, this remains the most genuine overture by Bill or Hillary Clinton toward a truly middle-ground initiative, widely heralded by moderate Democrats and Republicans across the board. Hillary's closest allies on the far left, from Patricia Ireland at the National Organization for Women to Marian Wright Edelman, did not share in this enthusiasm.

Edelman's Children's Defense Fund predicted nothing short of social Armageddon, with Edelman tapping into Hillary's social gospel to impugn the morality of the welfare initiative. Edelman wrote an open letter to the *Washington Post*, calling upon Bill Clinton's "unwavering moral leadership" to oppose the "tragic" and "morally and practically indefensible" welfare-reform package, "which will make more children poor and sick." The bills, said a righteous Edelman, constituted "fatally flawed, callous, anti-child assaults" upon "voiceless children," and would "eviscerate the moral compact between the nation and its children and the poor." Clinton's response would serve as the "defining moral litmus test" of his presidency.[1]

Throughout the 1970s and 1980s, Hillary had joined Edelman in accusing politicians who disagreed with them on poverty policy as being heartless, immoral, and against children. Now Hillary and her husband were on the receiving end. She was feeling what it was like to have someone from the religious left target one's policies as "anti-child" and "anti-Christian," simply because of a reasonable disagreement over *means* to an end, not ends. Wrote Edelman to Hillary's husband: "Do you think the Old Testament prophets Isaiah, Micah, and Amos—or Jesus Christ—would support such policies? . . . There is an even higher precedent that we profess to follow in our Judeo-Christian nation. The Old Testament prophets and the New Testament Messiah made plain God's mandate to protect the poor and the weak and the young. The Senate and House welfare bills do not meet this test."

It was a bracing display of moral arrogance by Edelman. Sure, Jesus wanted Christians to help the poor, as Christian Republicans and Christian Democrats knew, but nowhere in the Gospels did the Messiah weigh in on whether he preferred centralizing or decentralizing Medicaid. Edelman, however, was certain that Jesus did not like block grants. Meanwhile, the *Washington Post*—which usually held a strict line of separation between church and state—gave her the platform to use her faith to question these policies.

Bill Clinton signed the bills. In response, Edelman's husband and Hillary's dear friend, Peter, resigned his post in the Department of Health and Human Services, saying this was "the worst thing Bill Clinton had done."[2] Contrary to Edelman's predictions, the 1995 welfare reform proved an enormous success—maybe the greatest domestic achievement of Clinton's presidency—continuing a government safety net for the poor while weaning millions from continued federal dependency.

Nonetheless, this meant that 1995 was another difficult year for Hillary. In the previous year, the likes of Edelman had rejoiced at the prospects of a near-nationalization of the health care industry— apparently knowing with certainty that Isaiah, Micah, Moses, and Jesus were proponents of socialized medicine as well—but the public had not. This meant Bill was heading to the center.

The year also marked the Clintons' twentieth wedding anniversary. "Like any other couple that has been together a long time," said Mrs. Clinton, "we have worked hard and endured our share of pain to make our marriage grow stronger and deeper."[3] While they would celebrate twenty years of marriage in the most public house in the country, 1995 was a year in which Hillary's spiritual views would become more of a national issue than in any year previous, causing a stir for herself and her husband's office.

Her first public show of faith that year was not unconventional, and included sentiments that the Christian right should have lauded. On February 2, 1995, the first lady offered remarks at the National Prayer Luncheon. She made a solid statement on how a public person's spiritual faith should not be restricted to events like prayer luncheons:

The last time I spoke in public about spirituality, around the time of my father's death, I was astonished to realize that there were many people [who insisted that] spirituality should be confined to events like this, and not brought out into the public arena. I was amused when one commentator wrote that my critics were divided between conservatives who suspect I did not mean what I said and liberals who feared that I did. And I have become accustomed over the past year to living between those kinds of poles and trying as best I can to navigate what is for many of us uncharted terrain. Because as my husband said this morning, freedom of religion does not mean, and should not mean, freedom from religion.[4]

Despite speeches like this, a problem for Hillary was that conservatives (and many moderates as well) were not taking her faith seriously, largely because her leftist past made them suspicious of her politics. Her health care initiative had not done much to bolster their opinion of her, either, and their general distrust of her husband was not doing her any favors. As a result, many of her overtures on the matter of her faith seemed to fall on deaf ears, as the people that she seemed most interested in winning over were precisely the constituency that evaded her. For conservative Christians, that skepticism was about to receive what seemed like clear vindication, as Mrs. Clinton delved into something that sent practicing Christians—and many others—into orbit.

New Age Gurus

As Hillary's involvement with administration policy receded, she began to take up new spiritual interests within the White House. It started when she began to invite spiritual advisers who were both in and outside the mainstream to consult with her and the president on matters of belief. This seems to have started as New Year's Day 1995 approached, when the Clintons—though it is not clear exactly whose idea this was—invited a group of popular self-help and motivational authors to Camp David. The goal was to take a close examination of the first half of Bill Clinton's first term and evaluate potential ideas for his reelection prospects. Normally, such a gathering would comprise political strategists and analysts, maybe a pundit or two, a journalist, perhaps a presidential historian; the Clintons, however, were seeking something more emotional than coldly political. So, on the weekend of December 30, 1994, to January 1, 1995, the Clintons met with, among others, self-described "peak performance coach" Anthony Robbins, who on late-night paid TV commercials exhorted individuals to "awake the giant within"; the more down-to-earth

Stephen Covey, author of *The Seven Habits of Highly Effective People*; and, New Age "love" guru Marianne Williamson, who had recently presided at Elizabeth Taylor's latest wedding.

Though the three names were leaked to the press, all the meeting's participants kept the details of the get-together private. Not leaked, however, were the names of the other two advisers at the meeting, who were protected maybe in part because one of the two is highly controversial and was quite influential to Hillary. Ultimately, the two were revealed in a blockbuster scoop by Watergate reporter Bob Woodward in his 1996 best-seller, *The Choice*.[5] One of the pair was Mary Catherine Bateson, an anthropology professor at George Mason University, just a few miles down the road, and daughter of the famous anthropologist Margaret Mead. She had written a book called *Composing a Life*, which profiled the lives of five different women on "non-traditional" paths, which had become one of Hillary's favorite books.

The other, and the most unusual of the group, was Jean Houston, a woman in her mid-fifties widely known for her work delving into altered consciousness, the spirit world, and psychic experiences. Houston's specialty was taking herself and her subjects back into past worlds, both real and mythical, connecting them to long-deceased individuals as a method of finding personal comfort, stability, and healing. Houston had done this for herself, concluding that her own archetypal predecessor was the Greek goddess of wisdom, Athena. Claiming to have had many lengthy discussions with Athena, Houston was said to wear a medallion of the mythological goddess around her neck, including during her sessions with the first lady.

Of all the personages in the Camp David coterie, Houston was the most outlandish. As Woodward noted, it was Houston in particular who "saw possibilities" in the Clintons' "extraordinary openness about their pain." While on the surface Hillary might have seemed less susceptible to Houston's brand of alternative spirituality than her husband, upon their first meeting, it was evident that Hillary was taken by this spiritual thinker. According to Bob Woodward, "Hillary and Houston clicked, especially during a discussion of how to use the office [of the first lady] for the betterment of society." Through their dialogue, Houston had come to the conclusion that Hillary was personally carrying the burden of five thousand years of women being subservient to men; this was her cross to bear. Now, affirmed Houston, history was at a turning point, on the brink of genuine gender equality, and it was Hillary alone who could turn the tide—another Joan of Arc. Houston reportedly told Hillary that, next to Joan of Arc, she was there on the front line as arguably the most pivotal woman in all of human history. But she was a victim, a sufferer of bitter, unjustified personal attack; she was, said Houston, like Mozart, history's greatest composer, but with his hands cut off.[6]

"Though Houston did not articulate the image," wrote Bob Woodward, "she felt that Hillary was going through a female crucifixion. She had perhaps never seen such a vulnerable person, but also one who was so available to new ideas and solutions."[7] Nonetheless, said Woodward, Houston told Hillary she would prevail. She must persevere, as the new possibilities for the world's women were too much for her to cast aside. Her moment would arrive when she could accept and grasp that "fullness" of self.

In addition to these suggestions, Houston helped Hillary identify the means for fulfilling her global, millennial potential: She should proceed with the book on child care that she had been contemplating (the book that would eventually become Hillary's best-seller *It Takes a Village*), and take part in a UN conference on women in 1995, specifically, the Fourth World Conference on Women in—of all places— Beijing, to be held September 4–15. Recommending that Hillary attend the conference in Beijing was interesting advice, since Bill's political consultants had specifically advised Hillary to stay out of the conference. Her first-term entanglements in political controversies had already hurt her husband's ability to appear as a New Democrat in the 1996 race; now going to the UN conference, where international activists were campaigning for worldwide abortion rights, might yet again yank her husband and his presidency back to the left.

In the face of the campaign's political advisers, Hillary refused to be derailed from the path that Houston had set her on. According to Woodward, Houston wrote a strong letter to Hillary, underscoring her "obligation and burden, on behalf of all women, to go [to Beijing] and speak out." She enclosed letters from two friends likewise prodding the first lady to undertake this crusade on behalf of the world's embattled women: Until women had access to birth control and abortion, they were not free. It was a stirring text that struck at the heart of Hillary's most passionate spiritual and earthly causes; Woodward reported that Hillary said she cried when she read the letters.

However, rather than easing Hillary's burden, it seemed as if Houston was adding to it. Hillary's involvement with the conference in Beijing would most likely have a negative impact on her husband's campaign, especially given the fiasco with Vice President Gore at the Cairo conference the year before. The Vatican in particular was gearing up for another fight with the Clinton administration. Recall that the year before, the official papal spokesman, Joaquin Navarro-Valls, had taken the unusual step of singling out Gore by name; it was possible that involvement by Hillary could prompt the pope's spokesman to mention "Mrs. Clinton" next.

Over these reservations from the political operatives within Bill's campaign, Hillary went to Beijing and on September 5 gave one of the most memorable speeches of her career. This was a careful address, at times redundant but presumably for the purpose of reaffirming the theme, which was that "human rights are women's rights" and "women's rights are human rights." The only time that she used the word "abortion" was to denounce the host Chinese government for forcing women to have abortions against their will. That condemnation in Beijing demonstrated Hillary's ability to venture headfirst into confrontation, and it was such a powerful gesture that it muted some of her critics.[8]

Others, however, were not so easily satisfied with her remarks. While Hillary did not actually use the word "abortion" elsewhere in the talk, she used substitute phrases like "family planning." Most alarming to her detractors, the first lady affirmed an international "right to determine freely the number and spacing of the children" that a woman desires, implying without directly stating that abortion was a basic "human right"; in fact, ZENIT, the international news agency that covers the Holy See, later reported unambiguously that she had called abortion a "human right."[9]

Here Clinton had employed a clever rhetorical tool that would become her hallmark when discussing abortion in the years ahead as she pared down her language to reflect a more centrist stance on the issue. While her speech had all the markings of a typical Hillary speech, she created a linguistic sleight of hand by not mentioning abortion directly (and then only in a negative context), and that allowed the speech to rise above the anticipated criticism. The end result permitted the first lady to avoid the tidal wave of negative attention that Bill's advisers had feared and portrayed Hillary as someone who was moving to the center on this issue. She closed her speech by wishing "God's blessings" on all of those at the conference.

Although the Beijing trip and Houston's role in it appeared to be a success, it elicited a host of questions over the extent to which Hillary had been "taken in" by Houston. Her open disregard for the recommendations of Bill's political advisers demonstrated her passion for the cause, but it also provided yet another example of how Hillary—when given the right mentor—was capable of bending in ways that surprised even her closest allies. Houston's beliefs involved practices that required a suspension of Hillary's more conventional faith. Indeed, it seemed that Hillary may have latched on to Houston's views to fill the void that existed after her husband's political advisers stripped away her power. Whatever the reasons, such practices were not things that would be accepted in most houses of worship.

The Eleanor Incident

The success of the China trip reinforced Houston's rapidly escalating role as Hillary's closest spiritual adviser. At one point, says Bob Woodward, in October and November 1995, after Beijing, Jean Houston "virtually moved into the White House residence" for several days at a time to help Hillary—an encourager, an idea person, a "spirit raiser."[10] Despite the bond that Hillary seemed to be forming with Houston, this new friendship would be tested and eventually broken as less public facts about Houston's spiritual advisement began to surface.

A few months before the China trip, in the spring of 1995, Houston caught a glimpse of the large picture of Eleanor Roosevelt in Hillary's office. Houston, too, was a big fan of the former first lady, and as a teen had met Eleanor several times. As she related those encounters to Hillary, the two formed a strong connection over their shared love of the legendary woman. They talked about Roosevelt's struggles on behalf on social justice—for the poor, for victims of racism, sexism, classism. "*God,* this is really a serious Eleanor Roosevelt aficionado!" exclaimed Houston of Hillary.[11]

In retrospect, it seems surprising that it took Houston so long to recognize Hillary's connection to Eleanor; after all, Hillary had not been silent on her identification with FDR's first lady. Hillary routinely invoked Eleanor's name and sometimes referred to her in speeches. On February 21, 1993, Hillary had spoken at a fund-raiser for a memorial statue for Eleanor, where she went so far as to mention imaginary discussions between her and the dead wife of FDR. "I thought about all the conversations I've had in my head with Mrs. Roosevelt this year, one of the saving graces that I have hung on to for dear life." Rather innocently, she told the dinner crowd that she had asked Eleanor questions like, "How did you put up with this?" and "How did you go on day to day, with all the attacks and criticisms that would be hurled your way?"[12] This imagery did not strike most in the audience that day as strange or worrisome, especially since the current first lady, more than any before or since, brought comparisons to Eleanor Roosevelt. To many people, it was fitting that she identified with Eleanor; like Eleanor, she was strong-willed and determined, and their politics were obviously similar. Like Eleanor, she possessed a tenacious grasp on the ills of society and the steps that she believed necessary to remedy them.

Houston was willing to take these conversations much further, using them as an opening to another world. In Woodward's account, he describes how, during a visit to the White House in early April 1995, Houston proposed that Mrs. Clinton "search further and dig deeper" for her connections to Mrs. Roosevelt. As Woodward suggests, this was dangerous ground, since Houston and her work delved into "spirits and other worlds, put people into trances and used hypnosis, and because in the 1960s she had conducted experiments with LSD." With Hillary and her husband, however, she had "tried to be careful . . . intentionally avoiding any of those techniques." No LSD, mercifully.

Even without the drugs, this alternative spirituality seemed like a leap for the self-described "old-fashioned Methodist." These ideas that Hillary seemed to be supporting were far from traditional and quite out of step with Hugh Rodham's roots. At least part of Hillary's openness to Houston's unconventional background may have come from the fact that she and Chelsea had just returned from a two-week-long trip to the Far East, the home of Eastern mysticism. This, too, may have been a deliberate move on the part of Houston, who according to Woodward had fervently encouraged Hillary to take the trip.

One day in April 1995, Houston, along with Bateson, arranged to take Hillary to the solarium atop the White House, where they sat around a circular table. The solarium was reportedly Hillary's place of choice for important meetings, a room she had personally redecorated. They were joined by a number of members of the first lady's staff, though it is not clear if the staffers were there as nervous protectors or eager participants. One of them apparently tape-recorded the session, but the tapes have never been released.

As they were seated together at the table, Houston encouraged Hillary's thoughts about Eleanor. Woodward writes that Houston did everything she could to create a soothing atmosphere, one in which Hillary felt as comfortable as possible. Houston spoke openly to Hillary and tried "to create an atmosphere of mutual admiration." She called on the first lady to shut her eyes, eliminate her current surroundings, and envision herself in a room with Eleanor, a room where she was free to talk to the former first lady about whatever she wanted. Hillary did so, conjuring an image of a slightly frumpy but smiling Eleanor. Woodward's story continued:

> Hillary addressed Eleanor, focusing on her predecessor's fierceness and determination, her advocacy on behalf of people in need. Hillary continued to address Eleanor, discussing the obstacles, the criticism, the loneliness the former First Lady felt. Her identification with Mrs. Roosevelt was intense and personal. They were members of an exclusive club of women who could comprehend the complexity, the ambiguity of their position. It's hard, Hillary said. Why was there such a need in people to put other people down? . . . I was misunderstood, Hillary replied, her eyes still shut, speaking as Mrs. Roosevelt. You have to do what you think is right, she continued. It was crucial to set a course and hold to it.[13]

Houston was deeply moved. According to Woodward, she concluded that Hillary was facing "much greater toxicity and negativity" from her detractors, more so than Eleanor had. The current first lady "needed to unleash" her "potential." It was through adversity that she needed to find the "seeds of growth and transformation." Only then, wrote Woodward in his account, would it be possible for Hillary to "inherit" from the historical figure of Eleanor Roosevelt, "and to achieve self-healing."

But the session did not end with Eleanor. Houston next connected Hillary to Mahatma Gandhi. The first lady seized the reins, expressing reverence and respect for the Hindu leader's life and works. She found empathy with his persecution; he, too, like her, had been "profoundly misunderstood, when all he wanted to do was to help others and make peace."

Then Houston went higher, proposing that Hillary talk with no less than Jesus Christ—"the epitome of the wounded, betrayed and isolated," wrote Woodward, "and Houston liked to quote his words in the Gospel of Thomas, 'What you have within you that you express will save you, and what you have within you that you do not express will destroy you.'" Here, at last, the girl from Park Ridge Methodist drew a line: She said that talking to Jesus would be too personal in this setting. Amen.

They stopped the session after an hour. Elsewhere in the building, coincidentally, something had nauseated Chelsea, and she called for her mother. Hillary left to tend to her daughter.[14]

Nevertheless, Houston and Bateson—Bateson, ever the anthropologist, had been an observer rather than a participant in the Hillary-Eleanor session—graciously offered their services for future encounters, and the relationship did not end. Woodward reported that later Houston and Bateson gave "a kind of seminar" for the first lady and her staff at an "evening summer barbecue" at the home of one of Hillary's staffers. Hillary continued her meetings and "in-depth discussions" with Houston and Bateson regarding the parallels between her life and Eleanor Roosevelt's. There was much more to the relationship; they became buddies, a sort of "old girls' club," in the words of Houston.[15]

Some of these sessions became so powerful and important to Mrs. Clinton that Maggie Williams, chief of staff to the first lady, said after one occasion with Houston, "Oh, Hillary's ticking. She's had her 'Jean fix.'" In this tumultuous period, reported Woodward, Hillary had ten or eleven "confidants," but Houston was the most dramatic.[16]

The Fallout

As more than one observer has noted, the revelation of the Eleanor exercise in Woodward's book *The Choice* was a major embarrassment to a woman who prided herself on projecting intellectual strength.[17] Much of the country was blindsided by the book's revelations, and many Democrats—both supporters and critics—were astounded by the alternative beliefs that this devout Methodist had been engaging in. *Quoting* Eleanor was great, but *communing* with her "spirit" was something else entirely. Pundits ridiculed Hillary, saying that she was "channeling" Eleanor Roosevelt and holding a "séance" with Eleanor's ghost—allegations that the first lady and her staff vehemently denied.

Yet these suspicions were not unmerited. Both Houston and her husband, Robert E. L. Masters, had participated in channeling countless times and had placed innumerable people under hypnosis at their home. Masters has had his patients channel the Egyptian god Sekhmet. In her book *Public Like a Frog: Entering the Lives of Three Great Americans*, Houston introduced three individuals that she said were available to be contacted through a trance or altered state of consciousness: Thomas Jefferson, Emily Dickinson, and Helen Keller. Somewhere along the line, Eleanor Roosevelt also presumably made herself available.

According to Jon Klimo, an expert on channeling, there are different forms of the practice, ranging from full-trance channeling to sleep channeling, dream channeling, light-trance channeling, clairaudient channeling, clairvoyant channeling, open channeling, and physical channeling, among others. Some of these involve the use of

Ouija boards, while others manifest themselves in scary forms like levitation and voice alteration. Among them, clairaudient channeling sounds closest to what Hillary was reportedly doing with Houston; it involves relaxing oneself in either a fully conscious or mildly altered state of consciousness and then listening to one's "innerself."[18]

Hillary was not, as far as we know, levitating above a table in the White House.

Nonetheless, Hillary's involvement with the practice had her treading on tenuous spiritual ground. Deeper forms of channeling to supposed spirits of deceased figures are widely condemned throughout all of Christianity, including branches that view the practice as bordering on the demonic. The more than thirty thousand Protestant denominations, not to mention the countless independent Protestant churches, disagree on endless doctrines, but are virtually unanimous in this condemnation, because the Bible itself is so clear on the matter. Certainly, mainstream Methodism does not condone such behavior. Based on Woodward's description, the Houston sessions with Hillary never ventured into the extreme forms of channeling, but they held that potential, and they seemed an odd choice for someone who had never appeared prone to such strange displays of faith. Whether it was thanks to smart politics, her conventional Methodist upbringing, or plain common sense, Hillary honored boundaries and avoided being swept up in the spiritual turbulence that seemed to follow Houston. She never went completely over the edge.

Though Hillary had kept herself in check to a degree, her behavior was unprecedented, and many in America from both the right and the left had a hard time overlooking the facts. In the wake of this episode, Hillary's legion of detractors seemed to multiply. Criticisms flew from both sides of the aisle as everyone tried to make sense of how America's first lady, the president's wife, had become mixed up in something that seemed so diametrically opposed to her stated religious beliefs.

To understand the basic rationale for her actions, many Americans needed to look back to the events of the most recent years. The level of scrutiny that was applied to her in the wake of her ill-fated health care proposal had reached remarkable proportions, and while she had never been popular with Americans, many vilified her as never before. Whereas in years past she might have turned to the church to help her navigate this difficult time, for whatever reason she chose a new venue in Houston. But instead of offering support and comfort through this difficult period, Houston's escapades only made Hillary a subject of greater scorn and derision. It was a difficult if not paradoxical situation that Hillary found herself in, but it was one that she herself had created.

Ultimately, Hillary was forced to come clean, sheepishly admitting that she was guilty as charged. To that end, Gail Sheehy judged that Hillary handled the controversy "quite well," citing Hillary's formal statement of explanation, in which the first lady described the meetings with Houston and others as an "intellectual exercise," adding, "The bottom line is, I have no spiritual advisers or any other alternatives to my deeply held Methodist faith and traditions on which I have relied since childhood." Hillary said of Marianne Williamson specifically: "She is neither my guru nor my spiritual adviser. She is a political supporter who has an intriguing view about popular culture."[19] Mrs. Clinton also tried to joke about these incidents, saying in a speech: "I have just had an imaginary talk with First Lady Roosevelt, and she thinks this is a terrific idea."[20]

Though Hillary tried to poke fun at herself and turn the story into a punch line, the truth was that she had not been forthcoming about the incident. Her acknowledgments were hardly a trip to the confessional—for that, America had to rely on Bob Woodward. She had engaged in some bizarre, spiritually loose behavior that forced many Americans to think anew about her position in the White House and her role in the Oval Office. If it had not been for Woodward's digging, the extent of this behavior would never have been disclosed.

The United Methodist General Conference

In April 1996, Hillary began her public effort to get back on track spiritually, returning to the road of conventional Christianity—or, at least, such was the image she and her staff were eager to convey. That month she got some help from her friends when she addressed the United Methodist leadership, who had evidently lent a hand to a fellow believer in her time of need.[21] On April 24, 1996, the first lady delivered the keynote speech to the annual United Methodist General Conference, a talk that included no references to Eleanor or even to Michael Lerner.[22] Like many of Hillary's religious speeches, it made no mention of abortion, and similarly, Bishop Richard B. Wilke, for whom Hillary had done legal work back in Arkansas, failed to bring up Hillary's stance on the issue.

Despite these omissions, this was one of Hillary's finest religious speeches. She began by talking about the church of her youth, about her mother and Sunday school, about Vacation Bible School, and even about some of Chelsea's current experiences at Foundry United Methodist Church. All of that laid the groundwork for the theme of the speech: Christians, insisted Hillary, must "put into action what we believe" in all spheres of life, including the public sphere, where secular liberals insist on separating religion from the policies and practices of the state. She was "heartened" to learn that the Methodist Council of Bishops had renewed its call "to make the welfare of children a top priority." "Children need us," said Mrs. Clinton. "They are not rugged individualists. They depend, first and foremost, on their parents, who bear the primary responsibility for their upbringing." She cited Jesus Christ as the chief motivation in her government-based health care ministry to children, stating:

> We know the biblical admonitions about caring for each other. We know so well what Jesus said to his disciples in Mark, holding a small child in his arms, that whoever welcomes one

such child in my name, welcomes me, and whoever welcomes me, welcomes not me, but the one who sends me. If we could only keep that in mind, and see in every child's face that faithful hopefulness. Take the image we have of Jesus—I can remember so clearly walking up the stairs so many times to my Sunday School class, and seeing that picture that is in so many Methodist churches, of Jesus as the Shepherd. Taking that face and transposing it onto the face of every child we see, then we would ask ourselves, "Would I turn that child away from the health care that child needs?"

In this one passage, she brought her two predominant interests—health care and children—together under the umbrella of religion, in a telling explanation of her motivations for universal health care coverage. With this speech, Hillary laid bare how her passion for children and her passion for health care were both rooted in her strong belief in Jesus's teachings. It was an important display of the relationship between her private Christianity and her public policy, especially since many of her political counterparts often downplayed the role that religion should play in motivating and selecting policy. Whereas so often she, like her husband, seemed to base the origins for her stances in politics, here she offered a transparent explanation for how she came to believe in the value of nationalized health care: God had led her there.

Equally significant, the first lady cited official church teaching as an influence in her actions: "For me, the Social Principles of the Methodist Church have been as much a description of our history as a prod for my future actions." The Methodist Church and its principles, said the future senator, provided direction when it came to families, schools, and in "policies" to be developed for "each" child. "We continue in this church to answer John Wesley's call to provide for the educational, health, and spiritual needs of children."

Hillary then went on to list a number of areas in which she was "proud" that the church "had been a leader," citing the quality of education, the expansion of "comprehensive" health care, curbing smoking among young people, promoting "parental responsibility," and strengthening marriages. She then made a statement that, again, would be welcomed by conservative Christians but would make secular Democrats cringe: Hillary Rodham Clinton found kinship in John Wesley's words that "the world is my parish"—"and if that be the case, then I am optimistic." Turning once again to the children, she spoke about her book *It Takes a Village*, pointing to the chapter "Children Are Born Believers." "I feel so strongly," said Mrs. Clinton, "that we owe our children a chance for them to have a spiritual life." She made it clear—perhaps clearer than she had stated publicly in years—that all American parents owe their children an opportunity to become children of God.

She concluded by saying she was grateful for her Methodist upbringing and said, "In the name of Jesus Christ, we are called to work within our diversity, while exercising patience and forbearance with one another." She spoke of humility and patience, and concluded with a plea to make "common cause" with others who seek "personal salvation" and share a commitment "for the work we must do in this world."

It was a powerful speech that manifested Hillary's intent to make a definitive break from the spiritual confusion of the previous year. Here was a public embrace of Christianity unlike any she had undertaken during her time in the White House, one that showed the direction of her spiritual compass and the focus for the coming years. Nevertheless, her comments at the Methodist Conference and other similar functions brought a new question into light: Were her speeches simply products of election-year politics? With her husband in the midst of a campaign and her reputation still suffering from the revelations about Jean Houston, it remained ambiguous whether her newfound fervor was the result of a genuine reevaluation and investment in her faith, or whether her revised emphasis on her faith was designed to help her husband pick up ballots at the voting booth.

Once again both Hillary's supporters and her detractors found themselves coming across the same set of questions that had long plagued her faith, as people tried to understand just how genuine were her professions of belief. On June 10, 1996, she appeared on *Live with Regis & Kathie Lee*, cohosted by Regis Philbin and Kathie Lee Gifford, both conservative Christians. There she stated: "You have to fight this feeling of hopelessness and helplessness in your own life as well as in the lives of people around you. . . . I'm blessed with the kind of religious faith and upbringing that has given me a lot that I can fall back on."[23]

And yet despite this appearance and others like it, her actions during 1996 revealed a crucial contradiction: During the previous year, she had relied on nontraditional spiritual techniques to help fight feelings of hopelessness and helplessness. In the face of public anguish unlike any she had suffered before, she sought reassurance from alternate means; she sought answers from a new spiritual realm. While she could have chosen to "fall back" on the Foundry Methodists for support during this period, she experimented with something new, a faith beyond Methodism whose practices were completely foreign.

In hindsight, there appears to be an element of desperation about Hillary's decision to join the ranks of Jean Houston. Certainly when juxtaposed with her subsequent speeches and statements in 1996 and beyond, it appears that her time with Houston was a temporary detour, a spiritual misstep along an otherwise conventional road of faith. Ultimately, the importance of Jean Houston lies less in the overblown facts of what may or may not have transpired at these "sessions" with Eleanor, and more in the plain truth that Hillary had not relied exclusively on her church, family, and friends, to see her through these difficult times. Something about the Methodism on which she had relied for her entire life did not suffice in this instance, and as such, she left it behind (albeit temporarily) so that she could pursue an alternate form of solace.

Though her faith returned to the fold, her angst and frustration as first lady would only magnify. While she and her husband would survive the election intact, with another four years promised ahead, the next crisis would be the most awful yet, and it would test the moorings of her faith and the bonds of her marriage like never before.

CHAPTER 11

Surviving the Second Term

In November 1996, Bill handily won reelection, holding off an uninspiring challenge from Republican Bob Dole. The Clinton-Gore ticket beat the Dole-Kemp ticket 49 percent to 41 percent. Clinton actually lost men, 45 percent to 44 percent, but swept women in a landslide, 54 percent to 38 percent.[1] Two months later, he was inaugurated a second time, as feelings of hope and triumph pervaded the White House.

Bill emerged from the election with a stellar approval rating, a wave that he rode for much of the next year. The economy remained robust, rivaling the growth of the Reagan years, and fully recovered from the intervening Bush recession of 1990–1991. In fact, economically, Bill looked like he might be one-upping Ronald Reagan, as the ghost that had hounded the Reagan presidency—and every presidency since FDR's New Deal—the deficit, was slowly disappearing under an onslaught of rising tax revenues produced by the booming economy and disappearance of Cold War and Gulf War military budgets. On top of that, Clinton capitalized on the success of the rapidly maturing dotcom industry, which seemed to be manufacturing capital and wealth on an unparalleled scale.

Amid these triumphs, however, other events were stirring. There were the usual charges of Bill's caddish misbehavior, but the list seemed to be growing with names from Bill's present as well as his distant past. Of course, some were new, with sudden allegations of Oval Office fondling and groping by Democratic Party campaign worker Kathleen Willey, of past sexual harassment by Arkansas native Paula Jones, and yet more names from the old days: Elizabeth Ward Gracen, Sally Perdue, Dolly Kyle Browning, Gennifer Flowers. There would even be the most explosive claim of them all when Lisa Myers and NBC News went public with Juannita Broaddrick's rape allegations against former Arkansas Attorney General Bill Clinton; the mainstream press followed, including a detailed report on the front page of the *Washington Post*.[2] Despite these and other allegations, Clinton evaded any real scrutiny of his questionable behavior.

That changed in January 1998, when the forty-second president was accused of carrying on a not-so-romantic physical relationship with no less than a White House intern, a California girl named Monica Lewinsky. The affair had allegedly gone on for quite some time, and amid ever-present allegations of other Clinton sexual improprieties, this one was about to be confirmed, and to explode, eventually, all the way to a grand jury.

Bill had initially figured he could easily refute these claims of wrongdoing as he had all the others. Matt Drudge and other sources first broke the story on January 7, 1998, just as the first couple was happily announcing a new day care initiative about which Hillary was hopeful and proud. Joyce Milton called it Hillary's last truly happy day in the White House.

Two weeks later, on January 21, 1998, the story mushroomed, as the exceptionally fair Jim Lehrer of PBS's *The NewsHour* intrepidly asked the president about the Lewinsky matter. In that interview, Clinton said a number of times, "There is no improper relationship. . . . There is not a sexual relationship." But the issue intensified on Web sites and talk-radio shows, before coming to a head on January 26, 1998, when Bill stood before the press, conjured up a stern look of earnestness, shook his finger, and said believably, "I did not have sexual relations with that woman, Miss Lewinsky."

This denial, too, was well done. But it was a lie, a zinger of a falsehood that even his wife had bought. The next day, January 27, Hillary Rodham Clinton uttered words she would come to regret, claiming there was a "vast right-wing conspiracy" out to get her husband. In the general sweep of things, she was correct—there were untold numbers of right-wingers out to get Bill; this time, however, the dirt they had dug up happened to be true.[3] According to *Newsweek*, Hillary "honestly believed" that her husband had changed his ways once they arrived in Washington—as did tens of millions of other liberals and Democrats who voted for the man. When the Lewinsky story appeared, she had told friends that it was false—a product of the same vicious falsehoods as before.[4]

Unlike previous rumors, this one carried evidence. In July, the FBI began doing tests on semen stains on a blue dress worn by the college girl during one of her encounters with the president in the Oval Office. The evidence would later confirm his sexual involvement with the young woman.

"A Lapse in Judgment"

The bombshell hit the public on August 17, 1998, with Bill giving a public admission in a prime-time speech on national television, and thus starting a nightmare for Hillary Rodham Clinton. The speech was hardly a tell-all. Bill Clinton was master of the parsed word. The way he phrased his admission, conceding a "lapse in judgment," could have allowed for merely a single sexual act. That was all he needed to admit, since there was only one verifiable stain. Moreover, he used the speech to take the offensive against Kenneth Starr, the official independent counsel demanded by Congress and authorized by the Clinton Justice Department to investigate whether Clinton had committed perjury.

Bill seemed more frustrated with Starr than with himself. David Maraniss observed that the speech was angrier than he expected, but was also "vintage" Bill Clinton.[5] Presidential historian Michael Beschloss said that he "hated to admit it," but the speech sounded like Richard Nixon in the middle of Watergate, in denial and not taking full responsibility, blaming others. Former close Clinton aide George Stephanopoulos agreed, stating that the president did not satisfy or end the questions, quite the contrary.[6]

Historical framework aside, it was apparent that his was a most reluctant confession. As former Democratic speechwriter turned pundit Chris Matthews put it, Bill Clinton "didn't decide to tell the truth, he got caught"—literally, one might say, with his pants down. Yes, chimed in legal analyst Stuart Taylor, Clinton was "a fundamentally dishonest man," who "cannot be trusted."[7]

In truth, President Clinton did say that he was "solely and completely responsible," though that stand-up statement seemed to fall by the wayside as he tore into Starr. He did admit, "I misled people, including my wife. I deeply regret that." Hillary first learned the truth from the lawyers, not her husband, who told her on August 13. She then began absorbing the treacherous details that were published widely in the press.

While Starr's behavior during the investigation has been debated and discussed in full by people on both the right and the left, it is worth noting that even the *New York Times* editorial board acknowledged that Starr was doing his appointed job: to confirm whether Bill Clinton had committed perjury and obstructed justice. This meant looking into details of Clinton's life, which meant, as Hillary knew, making unsavory discoveries of lurid private behavior and attempts to cover up the litany of transgressions—this time in ways that seemed to constitute federal criminal offenses. As the *Washington Post* noted in an editorial, "even when [Mr. Starr's] excesses are stripped away, the case he has presented is serious, while Mr. Clinton's current defense is contemptible." The *Post* stated that Clinton's behavior was "at the margins of impeachability," acknowledging: "There is . . . ample evidence in Mr. Starr's report of presidential conduct that Congress could deem grounds for impeachment."

Yes, Starr had reported some tawdry personal material. Yet Bill Clinton, publicly and under oath, had denied the entire relationship with Lewinsky. As the hired independent counsel, as the official government prosecutor, Starr had a legal and ethical duty to find and report the details that he learned existed. Had he not done so, he, too, would have been guilty of negligence. If he had muffled the information he was receiving, he would have been guilty of a cover-up. In the introduction to the report, Starr apologized to the president for presenting the salacious material, but stated clearly that he had no alternative but to incorporate it. To this day, Bill Clinton remains angry at Starr. As Starr noted, it would be up to others, namely the U.S. Congress, to decide what to do next.

When Starr's fact-finding mission finally became public knowledge, the story was lurid and the public was captivated. There had been eighteen months of gifts and at least a half-dozen sexual encounters between Bill and Monica. The first sexual encounter took place on the same day that Clinton signed a "Family Week" proclamation. On that November 15, 1995, Lewinsky, walking through the hall on her way to the ladies' room, spotted the president of the United States, and greeted him by lifting her jacket to show him the straps of her thong underwear—what the girl later winsomely described as a "little smile." Hillary's husband was intrigued, and invited the intern back to his office. That was where it all began.[8]

The public joined Hillary in learning the details of the spectacle, as the Starr Report became public in September. Among the most devastating items in the report, especially from the perspective of

Bill Clinton's faith and the public's perception of him as a religious (or nonreligious) man, was the tale of Clinton's sexual exploits with Monica on Easter Sunday, April 7, 1996, when Bill, who had spent the morning at church, had an afternoon liaison with Monica in the Oval Office. This liaison was perhaps the most scandalous of them all, having commenced in the hallway before moving to Bill's private study, whereupon the president of the United States received oral sex from the intern as he simultaneously conducted business over the telephone.

A few million Americans read about this Easter rendezvous in an excerpted section from the Starr Report in their weekly copy of *Time* magazine. A few million more read the lengthy account in the September 14, 1998, *New York Times*. Another million or so read about it in their morning *Washington Post*, or weekly *U.S. News & World Report*, or *Newsweek*, or wherever. The incident has hardly been forgotten—especially among observant Christians. Detractors and skeptics of Bill's faith never cease to bring it up, always eager to denounce Bill as a hypocrite. In fact, since the days he had attended Park Place Baptist Church as a lonely eight-year-old in the 1950s, up through Immanuel Baptist Church in the 1980s, Bill Clinton had strolled from church with a Bible tucked under his arm. All along, like all Christians, he remained a broken sinner, tormented by demons—as a child, it was the abuse in his household; as an adult, it was sexual temptation and dishonesty. Nonetheless, when the viewing public saw that Easter Sunday image, they thought of Monica. And they also thought of Mrs. Clinton.

More and more information followed, like the revelations of the many occasions of "phone sex," which the young Monica was excited to share with Barbara Walters in an exclusive interview on ABC. Monica told Barbara over and over about how the president was "so sensual. . . . I mean, he's just so sensual!" After Monica described the "fun" phone sex with the president, Barbara asked her if what she did for Hillary's husband was technically considered "sex," an important legal question, since many feared that the master of the parsed word would claim he had not committed perjury when he denied a "sexual" relationship with the intern. Monica said no. "Then what is it?" yelled an incredulous Walters. Said Monica: "Oh, that's just fooling around!"

Just when some people might have considered this a perverse form of comic relief, even as it belittled the stature of the Oval Office, Monica reminded the American public of the extent of her youth and her naiveté, stating that the president had confided in her that he "might be alone in three years," referring to the prospects of Hillary dumping him.[9] It was a stunning remark, and one that sent both Hillary's supporters and detractors clamoring for answers about how the first lady was responding to these developments. Suddenly, not only was the Monica tale true, but all the others—no matter how they once stretched the boundaries of credulity—seemed not only plausible but likely. Though Hillary had dealt with her husband's stories and cover-ups for years, the new realities that Monica revealed meant that little if anything Bill said could be trusted. It would be a large weight for any person to bear, but for a woman who lived her life in the very public scrutiny of the American media, it was downright impossible. With so many of Hillary's fundamental beliefs about her marriage publicly shaken to their core, the question was unavoidable: How would she weather the storm?

Monicagate

Depending upon whom one believes, or reads, Hillary's reaction to all this was either business as usual—since an unfaithful Bill was hardly a shock to her—anger at the political damage to her partner in power, or deep dismay over this betrayal of her sacred marital vows. Here, again, it is difficult to know what she truly felt in her heart, as the answer was known only to God and Hillary Clinton. Her friends told the press, "To say that Hillary is disappointed would be a gross understatement."[10]

But no matter how much the Clintons and their cronies tried to blame the crisis on Ken Starr for confirming it, or the likes of Matt Drudge for first reporting it, in truth Bill only had himself to blame. As Jonathan Alter wrote in *Newsweek*: "By committing sex acts with a 22-year-old subordinate in the White House, Bill Clinton made his marriage the nation's concern. In doing so, he victimized his wife all over again. . . . Hillary was only now bearing the full burdens of her bargain."[11] Yet, as Alter and others noted, her "disappointment" seemed as much political as personal. Wrote Alter: "She is said to believe that her husband betrayed not just her and Chelsea, but much of what they have worked together to build."[12] Indeed, the strange press release from the first lady's office referred to her husband in a political as well as a personal way, saying that she "is committed to her marriage" and "believes in this president and loves him very much."[13]

Nonetheless, said her friends, she turned "inward, to her spiritual side."[14] Her press secretary, Marsha Berry, stated, "Clearly, this is not the best day in Mrs. Clinton's life. . . . This is a time that she relies on her strong religious faith."[15] (In a June 2007 discussion of her faith, Hillary elaborated on Berry's statement, announcing unequivocally, "I'm not sure I would have gotten through it without my faith.")

There were in fact spiritual sources that Hillary tapped at this time, taking guidance from certain ministers. One such minister was civil rights leader Jesse Jackson, who invited himself to offer counseling even before Mrs. Clinton could pick up the telephone.[16] Jackson dashed to the White House to offer his services—and then dashed to the cameras to tell the media all about it, generously giving an exclusive on his role to *Newsweek* in an August 31, 1998, article.[17]

Jackson's offer of himself was not inappropriate, as he had often called Chelsea at college to pray with her, and had apparently started doing so back in January during the night of the Super Bowl.[18] During her talks with Jackson, Hillary mentioned that she felt Chelsea needed Jackson's counsel more than she did. At eighteen, Chelsea was not much younger than Monica; Chelsea was a college freshman, while Monica was the age of a college senior. So, on this August night, Jackson called Chelsea at the White House to ask if he could help. She fought back tears. "Things are really tough for my mom and dad right now. Would you be able to come over tonight and pray with us?"[19]

He said he would, right after he finished defending the president that evening on CNN. A government limousine picked him up from the studio to take him to the White House. A Secret Service agent escorted him to the private quarters on the second floor. Hillary and Chelsea were wearing sweats—casual attire for a Sunday evening. They met Jackson in the Center Hall. Jackson later described them as "devastated." With somber faces that, according to Jackson, had clearly been crying, they hugged him and then walked together to the Yellow Oval Room.[20]

It was 10:30 P.M. on a Sunday night that August, just before Clinton was preparing to give his testimony to Ken Starr and his lawyers. Jackson and Hillary and Chelsea sat in the living room and talked. Bill dropped by to say hello. After a few minutes of bantering, Hillary ordered him upstairs to finish preparing for his testimony—she was not losing sight of larger political priorities. According to Jackson, Hillary told the president sternly, coldly: "When we're done talking, then we'll come up and see you." Like a little boy listening to his mother, Bill did what he was told.[21]

When Bill left, Jackson cut loose. He was ready, and told Hillary and her daughter that he was reminded of "another First Family in crisis, in the very first Rose Garden," when Adam and Eve transgressed from God's will and sinned. He talked about King David, who, he said, was like Bill Clinton—talented, gifted, but tempted by and succumbed to the forbidden fruit. All of that wisdom, and David could not control himself around Bathsheba. Likewise, Samson, with all his strength, gave in to the temptations of the flesh when he saw Delilah.[22] The civil rights leader said that God had forgiven these great men "in their weakness," and that the Clinton family needed to do the same; they needed healing from the Lord. Surely, Jackson could relate to this because of his own well-documented marital difficulties; he, too, had been led into temptation and needed forgiveness.[23]

This was true, claimed Jackson, of some of the greatest leaders in American history. "This storm," he told mother and daughter, "is not an original storm." The key question, Jackson said, was how should they respond? They could panic and jump overboard, they could become irate and point fingers and turn on one another, or they could "apply [their] faith." He advised the last, that they hold on to their best hopes and weather the storm together until it passed.[24]

He told the women that he saw this as a trial: "one's faith is only truly put to the test when you are forced to walk through the storm. And that is what this is all about: faith. Faith and unconditional love."[25]

In his retelling of the incident to *Newsweek*, Jackson did not feel at liberty to relay their response, but, he said, "it was clear that Hillary and Chelsea understood."[26] He shared his observations not only on how Chelsea was coping—which Jackson said was pretty well—but also on Hillary. "With grace and strength," said Jackson of the first lady, "she was coping with a crisis affecting her husband, her daughter and the nation—and there was almost no one she could talk to about it."[27]

Reverend Jackson did not finish until midnight, when he joined Bill, still awake, in the private residence. According to accounts, Hillary followed him there. The reverend complained irreverently, "What's different here is that Ken Starr is able to play God with government funding"—a seemingly misguided analogy but one that led Hillary to let out her trademark whooping laugh. "Where did you get that line?"[28]

Another source says that Bill personally was not quite so amused at the line, stating gravely that the family needed God's healing. Before breaking up the meeting, Clinton asked Jackson to stay a little longer to talk one-on-one with Chelsea. "I think she's confused by the whole situation," the president said. "If you could just let her know that these things happen." At that point, one of the Clintons, Hillary, finally went off to bed, leaving Jackson alone for that heart-to-heart with her daughter.[29]

Counseling Bill

As the drama continued to play out, Bill, too, needed some special spiritual attention. All humans are sinners, and all Christians are Christians because they acknowledge their inherent selfishness, sinfulness, and need for redemption not possible in and of themselves. Bill certainly realized this now more than ever. Consequently, he sought help.

Philip Wogaman, the Methodist minister at Foundry, became part of what would become known as the "God Squad"—the trio of ministers that began counseling Bill in this difficult time. He was joined by fellow Protestants Tony Campolo, a popular liberal evangelical from eastern Pennsylvania, and Gordon MacDonald, of Grace Chapel in Lexington, Massachusetts.[30] Campolo, a kind man known for preaching forgiveness, was, like Wogaman, a liberal who agreed with Clinton's politics. MacDonald was probably the one member of the God Squad who could most relate to Clinton, as he himself had committed an extramarital affair with a member of his congregation and had been forced to leave his church. He wrote a book about his struggle, which Bill Clinton said he read twice.[31]

Since the role of the God Squad first became public knowledge, there has been a great deal of confusion over how these men were chosen. One source says that before they accepted the assignment, the three ministers asked if Hillary approved of the idea. According to this source, one aide replied incredulously, "Approve? Hell, it was Hillary's idea."[32] Though multiple sources cite Hillary as the impetus behind the formation of the God Squad, Bill's memoirs credit himself: "I had asked three pastors to counsel me at least once a month for an indefinite period," he wrote.[33]

In an interview for this book, Wogaman confirmed that Bill, not Hillary, was the impetus. Hillary did not approach him about counseling her husband; rather, said Wogaman, it was Bill's idea to approach him. Wogaman said the three ministers "would trade off" so that Bill could meet with one of them each week, though there were occasions when all three were present and counseled him. Wogaman said that this counseling lasted until the end of Clinton's presidency; they must have all concluded that this needed to be ongoing. Thus, Wogaman confirms, Bill, in addition to his regular church attendance, was meeting once a week with personal spiritual counselors, meaning that he was having at least two meaningful spiritual encounters a week, whether at Foundry, the White House, or during chapel services at Camp David.[34]

This was not necessarily stepped-up activity for Clinton. Even before the Lewinsky scandal, Campolo says that he and Clinton "would get together about once every five or six weeks for a couple of hours." After the scandal, said Campolo in an interview for this book, "I got together with him much more often." He adds, "I continued the relationship of intense counseling all through the rest of his presidency."[35]

Clearly sensing certain political needs as well, Clinton telephoned Campolo after the Lewinsky scandal broke to ask Campolo if Clinton could publicly announce that the minister would be doing "pastoral work" for him. This was "his initiative" to make this request, said Campolo, who quickly agreed to the president's request.[36]

Both Campolo and Wogaman were reserved about the nature of these post-Monica discussions. Wogaman responded that "much of it was obviously off the record and personal, but we discussed how important faith is in forming our lives." The minister said that through these experiences, he learned that Clinton's faith was "genuine," though Wogaman did carefully concede that "you can never entirely read into the heart of a person." Nonetheless, Wogaman says that he formed a "great opinion of him [Clinton]. He's very quick, he reads a lot, and retains it. He's a person of deep commitment who sought to relate faith to life and didn't always succeed."[37]

In his first session with the God Squad, Bill opened his Bible and read his favorite passage from Isaiah: "And I shall run and not grow weary."[38] Wogaman perceived that Clinton was "very familiar with the Bible." He never specifically asked Clinton whether he did any sort of daily devotional practices, but he was "confident" that Clinton prayed and read the Bible on a regular basis. "He appeared to me as a deeply connected spiritual person," says Wogaman. In their personal conversations, Wogaman said there would always be a "highlight of at least one scripture passage."[39]

Through his time with the president, Campolo came to many of the same conclusions as Wogaman, and additionally came to understand more specific elements of Bill's beliefs. During their sessions, Campolo learned that Clinton holds to an evangelical theology, affirms the doctrines of the Apostles' Creed, and "believes the Bible to be an infallible message from God." Like Wogaman, Campolo pointed to Clinton's commitment then and today to biblically inspired social justice—like Hillary: "[H]e is especially committed to living out the two thousand verses of Scripture which call upon us to respond to the needs of the poor," says Campolo. "Both in the presidency and since leaving the presidency, the verses concerning serving the poor have guided his life."[40]

According to one source, the trio was reportedly instructed not to invite Hillary to participate. "Mrs. Clinton," said one of the three, "did not want to be part of our counseling sessions, period."[41] Hillary knew that it was Bill who had the problem.

Nonetheless, as for Reverend Wogaman, he had more than Bill to consider at this trying time. After all, Hillary, as well as Chelsea, was a member of his church. Asked if during this period he ever spoke to Hillary and Chelsea about serious spiritual matters in relation to Bill and his extramarital activities, Wogaman said only, "That's personal."[42]

Regardless of the individual roles that were played, the entire family was calling upon their faith to help them make sense of this difficult time. Whereas just three years earlier, Hillary had taken consolation in the ideas of Jean Houston, now it seemed that her faith in God was one of the only things that could support her through this ordeal.

Because this was a marriage playing out on the most public of stages, their problems would have to be resolved in front of the whole world. Whatever conclusions they came to together would not be strictly between them and God. There was a third party involved as well: the American people. For better or for worse they had been pulled into this marriage, as Bill Clinton, ever the master politician, was not about to forget.

Forgiveness

While much of Bill's prayer and meditation on sin had been happening behind closed doors, as events progressed, his need for forgiveness took him to a public stage. A week and a half after his August 17 admission, speaking at an August 28 event in Oaks Bluff, Massachusetts, marking the thirty-fifth anniversary of Martin Luther King Jr.'s "I Have a Dream" speech, Bill spoke of the need to seek forgiveness: "it is important that we are able to forgive those we believe have wronged us, even as we ask for forgiveness from people we have wronged." This was a step up; he was at least talking forgiveness, but it was self-centered: He wanted forgiveness from Hillary, who would be among the "people we have wronged," and seemed to speaking of

Ken Starr when he spoke of forgiving "those we believe have wronged us."[43] He was still pointing the finger at Starr.

Yet as his speeches and his appearances continued, he began to turn inward, focusing on the role of his personal responsibility and the role of repentance. Indeed, he spoke of forgiveness constantly—in Moscow on September 2, in Dublin on September 4, in the afternoon and then the evening at separate stops in Florida on September 9. In Dublin, he had used the word "sorry."[44] Then, on September 11, 1998, Clinton gave what was much closer to a confession, inviting a large contingent of African-American religious leaders to the annual White House Prayer Breakfast for Religious Leaders for a kind of half-revival, half-Baptist-style cleansing. The assembled clung to the edge of their seats as Bill started by acknowledging that he had been on "quite a journey" over the last few weeks. He needed to "get to the end of this, to the rock-bottom truth of where I am and where we all are." He then made a real admission: "I agree with those who have said that in my first statement after I testified, I was not contrite enough. I don't think there is a fancy way to say that I have sinned."[45]

The president then expressed sorrow for all those he had hurt, from his family to Monica's family to his staff, and asked for forgiveness from everyone. He said he believed that "to be forgiven, more than sorrow is required." The first step, said Clinton, was the need for "genuine repentance" by the sinner, "a determination to change and to repair breaches of my own making," to repair "what my Bible calls a broken spirit." He proclaimed, "I have repented." That repentance, he conceded, needed to be "genuine and sustained, and if I can then maintain both a broken spirit and a strong heart, then good can come of this for our country, as well as for me and my family."

This personal purging was met with shouts of amen and hallelujah from those present, as Bill said it was now a "time for turning," to start "all over again." This is a "painful" process, said the sinner, which "means saying I am sorry. It means recognizing that we have the ability to change. These things are terribly hard to do. But unless we turn, we will be trapped forever in yesterday's ways." In this strikingly candid speech, Clinton then made a vow to these fellow Christian witnesses: He would continue to seek the spiritual counsel of the three pastors with whom he had been meeting, noting that they were assisting him in overcoming "the temptations that have conquered me."

This was not merely a gathering of Christian men, of promise keepers, Hillary was there also, as was her old pastor from First United Methodist in Little Rock, the Reverend Ed Matthews, who in the past had dealt with questions of Bill's womanizing. Matthews had not lost touch with Hillary, recalling one occasion when she returned to Little Rock for a service after moving to Washington; she alarmed the Secret Service that day when she rose to help pass the collection plate, as she had always done.[46] Those were better days. This one on September 11, 1998, was not.

In an interview a year later with Gail Sheehy, Matthews relayed Hillary's emotional reaction to her husband's witness that day. "She had tear-filled eyes," said Matthews, who had been sitting at the table next to Hillary and immediately went over and knelt next to her as soon as her husband left the podium. He said she whispered to him, "I never heard him say that before. And you know what we've been through." She and Matthews then talked about that forgiveness that her spouse was seeking. "You know that forgiveness has nothing to do with human logic," he told her. "Forgiveness has strictly to do with grace. And that's God's gift. I hope you are in touch with that. I'm praying that you are." Hillary then smiled and said, "I think I'm getting there."[47]

In 2005, Matthews talked about the incident again. Matthews reported that Bill and Hillary had talked about the Monica affair the night before, though he could not relay any details. That afternoon, the pastor recalled that it had "felt good" for him to be able to take

Hillary's hand at the luncheon and "let her cry." Asked if Hillary actually cried to him, Matthews clarified:

> No, she didn't cry about it. . . . She's not that kind of emotional person. I moved across the room and knelt down beside her and visited for a couple of minutes. . . . She talked about how they had met about it the evening before; I would describe her as in a state of distress. . . . She said to me, "How am I going to get through this?" I said, "Well, it depends on how well you stay in touch with forgiveness and how God provides it." She did this in a remarkable way. She understands that forgiveness is outside the human agenda. That conversation took place in those two minutes. I spoke to her about forgiveness.[48]

Matthews added, "Those of us who understand forgiveness know that it is outside of human economy—it is God's gift—there is no way we can forgive without Him." He says that Hillary grasps this concept from the recesses of her Wesleyan tradition.

Healing and forgiveness meant getting back to church. On September 20, 1998, for the first time since August 17, when Bill admitted to his momentary lapse with Monica Lewinsky, the Clintons returned to church, arriving for the 11 A.M. service at Foundry.

They had taken a thirty-four-day hiatus from their normal weekly churchgoing, perhaps figuring the public would view Bill as a hypocrite if he was seen walking in and out of church with a Bible tucked under his arm after admitting his inappropriate relationship with Lewinsky—a misperception often held by people who do not understand that this ever-present tendency to sin is what compels Christians to need to go to church. The Clintons had skipped two weeks of church while vacationing on Martha's Vineyard, and then again avoided services after their return to Washington.[49]

Apparently the healing process did indeed begin. In his memoirs, published in 2004, Bill said that he and Hillary "began a serious counseling program," which occurred one day a week for about a year. "For the first time in my life," wrote Bill, "I actually talked openly about feelings, experiences, and opinions about life, love, and the nature of relationships. I didn't like everything I learned about myself or my past." He said that in these "long counseling sessions" and their conversations about them afterward, he and Hillary "got to know each other again, beyond the work and ideas we shared and the child we adored. I had always loved her very much, but not always very well." He said he was grateful that she had been "brave" enough to participate in the counseling: "We were still each other's best friend, and I hoped we could save our marriage." Yet, in the meantime, he explained, "I was still sleeping on a couch"—a sofa in the small living room that adjoined their bedroom, where he was confined for at least the next two months or more. Eventually, said the president, he "got off the couch."[50]

In the end, the Clintons managed once again to keep their marriage together. Of course, a divorce of a sitting first couple was certainly out of the question, especially for these two partners.[51] And the first lady did extend forgiveness, later telling ABC's Barbara Walters that she had forgiven her husband's "sins of weakness." She had reached peace. "I don't think people know how strong your faith is," Walters remarked. "It must have helped you." Yes, Hillary replied. "I've relied on prayer . . . ultimately I had to get down on my knees and I had to pray and I had to look for answers that only could come to me."[52]

The Limits of Forgiveness

Though Bill and Hillary were each relying on the intertwined ideas of faith and forgiveness in ways they never had before, they each had limits to those feelings, and those limits stopped at Kenneth Starr.

For the remainder of 1998, people throughout the administration sought to portray Starr and the special prosecutor's office as being overrun by puritanical Christian fundamentalists who were committed to the demise of the president. Clinton supporter and Democratic strategist James Carville lampooned Starr for his allegedly unsophisticated, embarrassing practice of singing hymns and praying during jogs along a creek near his home. "He goes down by the Potomac and listens to hymns, as the cleansing water of the Potomac goes by, and we're going to wash all the Sodomites and fornicators out of town," screeched Carville. Likewise, Sidney Blumenthal, a White House aide, said that Starr was a "zealot on a mission derived from a higher authority," and made similar comments about the independent counsel's staff, such as Starr's chief deputy in Little Rock, W. Hickman Ewing, who was labeled a "religious fanatic."[53]

The public mockery of Starr's office was a regrettable component to the White House's counteroffensive during this time. In a telling display of spiritual double standards, the public face of the Clintons portrayed their faith in God as their great stabilizer in this time of turmoil, while behind the scenes their aides were quick to ridicule those on the opposing side who also relied on their spirituality to sustain them.

As Laura Ingraham—a political pundit and author of a book on Hillary—pointed out, the one voice who could have apologized for this behavior in an authoritative way was Hillary the devout Christian. Though she would not have needed to accept all of Starr's arguments, she at least could have become a forceful voice against staff members' caricaturing of Starr as an irrational fundamentalist. Yet she remained silent, allowing the stigma against belief to percolate throughout the administration's public face.[54]

The point is well taken. She could have risen above the mudslinging, setting a higher standard. The politics would have been brilliant. By then, however, the dogs were so ferocious and the vituperation so strong that cool heads were hard to find.

By countering the special prosecutor's office with accusations of Christian extremism, the Clinton staff offered the American public reasons to disavow Starr's conclusions. While the Clintons might have believed that this would help them dodge political heat, there was another unanticipated result that a moral Methodist like Hillary never could have anticipated: In their willingness to bend over backward to defend their president, and to dismiss Bill Clinton's behavior as merely a Republican attempt to smear the Clinton presidency, many liberals, and many feminists, made the situation even worse, searching for responses that further coarsened what even Hillary judged was an overly sexualized culture. In one defense of Bill Clinton, Nina Burleigh, who covered the White House for *Time*, wrote in *Mirabella* magazine, "I'd be happy to give him [oral sex] just to thank him for keeping abortion legal." Revealing more than a sliver of journalistic bias toward her subject, Burleigh added: "I think American women should be lining up with their presidential kneepads on, to show their gratitude for keeping the theocracy off our backs."[55]

Burleigh was far from the only person to dismiss Bill's behavior because of politics. Mike Nichols, journalist Diane Sawyer's husband, told the *Washington Post*: "If this century has taught us anything, it's that sexuality is uncontrollable. . . . We expect it to be tailored, controlled, changed . . . like a pet cat. But it's not going to happen." Thus, waxed Nichols of his president, "So our charismatic men with high energy . . . are beginning to say: 'The hell with this! I'm not going to get in a position where I'll be torn to bits every time I look at a female walking by.'"[56]

Likewise, actress Emma Thompson, who played Hillary in the film version of a book based on the 1992 Clinton campaign, *Primary Colors*, agreed that Clinton was paving a new frontier for "alpha males." She said of the Clintons: "powerful, brilliant women have been married to powerful, philandering men for hundreds of years. . . . You may marry an alpha male—that's what happens. This almost seems natural for any guy in power." Thompson found it "a very erotic situation."[57]

The problem, the cultural mavens were concluding, was not Bill Clinton, but the repressive Victorians complaining about Bill Clinton. So it went that the culture, not the president, was to blame for the response to his behavior. The culture, it seemed, merely needs another dose of 1960s counterculture to free itself from its taboo of adultery.

Many young people around the country—whom Mrs. Clinton and her husband had been trumpeting because of the recent decline in teen pregnancy (a trend for which the Clintons took credit)—suddenly found themselves confused and misled by Bill's suggestion that oral sex might not even be technically considered "sex." Dana Gresh, an evangelical author who specializes in outreach to young adults, alluded to how this attitude made inroads among conservative Christian teens, who began referring to oral sex as "Christian sex"—a kind of "non-sex" sexual activity that allegedly keeps one pure because it is not true, full intercourse.[58]

Finally, Bill Clinton's philandering led to ludicrous talk of Mrs. Clinton's husband suffering from a "sickness" called "sex addiction," a new, modern-day "disorder" particularly unique to Bill among adult males. Hugh Rodham's daughter would almost certainly have been the first to tell her husband that she did not want to hear any nonsense about an addiction. Likewise unfortunate was the defense that "everyone did it," meaning that "all" presidents had engaged in such behavior, which, of course, was equally ridiculous, and in the process demeaned the presidency even more. As Eleanor Roosevelt's husband once said: "The presidency is preeminently a place of moral leadership."[59]

These myriad justifications excusing the president's actions had the cumulative effect of overshadowing much of the actual debate over the morals and the behavior of the president. And the real losers in all this were the American people.

One person who vehemently disagreed with many of these rationalizations was Hillary, as Bob Woodward revealed in his 1999 book,

Shadow. There he reported that Hillary was "devastated," "humiliated," and she even deduced that this must be part of God's plan—it was her time to suffer, her cross to bear. "I've got to take this," she told a friend. "I have to take this punishment. I don't know why God has chosen this for me. But He has, and it will be revealed to me. God is doing this, and He knows the reason. There is some reason."[60]

In addition to the private weight she was feeling, she suddenly found herself compromised in her ability to make a difference on some of her most passionate issues. An embarrassing example of her newfound limitations occurred on July 16, 1998, during the buildup to the Starr Report, when Hillary addressed the 150th anniversary of the first Women's Rights Convention in Seneca Falls, New York. She asserted that women were still fighting for certain things, like better jobs and wages and "respect at home." At one point during the speech, some wag flew overhead in a private airplane with a giant banner that read, "Who's at home watching Bill?"

It was a difficult moment, one that demonstrated the obvious limitations of this new incarnation of public scrutiny.

Impeachment

"It is not in my hands," Bill Clinton told reporters. "[I]t is in the hands of Congress, and the people of this country—ultimately, in the hands of God."[61]

This was Bill's response to the question of how this whole fiasco would end. The House of Representatives laid down its verdict on Saturday, December 19, 1998: impeachment. The U.S. House of Representatives voted to impeach Hillary's husband by a vote of 228–206 on article one, "Did the president commit perjury before a grand jury?" and a vote of 221–212, on article three, "Did Bill Clinton obstruct justice?"

He had technically done both, the only debate being whether the perjury and obstruction involved a serious enough offense to merit impeachment. Democrats said no, but enough Republicans said yes, and on that day William Jefferson Clinton became the second president ever to be impeached and the first elected president to be impeached.[62]

The next big question was whether he would resign, as many urged him to do, from veteran *Washington Post* journalist David Broder to respected newspaper editorial boards like those of the *Cincinnati Enquirer*, *San Jose Mercury News*, *USA Today*, the *Philadelphia Inquirer*, the *Seattle Times*, and many others.[63] The pressure rose when Republican Speaker of the House Bob Livingston (R-La.), caught in his own sexual transgression, though one that did not involve legal breaches before a grand jury, resigned that same Saturday morning.

Bill, however, did not follow suit, opting instead to wait out the debate of his impeachment in the Senate. Despite the successful vote in the House of Representatives, the vote in the Senate was well short of the two-thirds majority required to convict and remove him from office. And so, with the conclusion of the Senate trial, Bill Clinton, his marriage, and his faith had weathered the storm. The cost of these events had been high, as Clinton had spent much of his political capital trying to clear his name. Though he would continue to make speeches and herald the increasingly vibrant economy, he no longer possessed the political strength that he once held dear.

Another Meeting with John Paul II

It never got worse for the Clintons than that December 1998. By the next month, the start of a new year, things began to get better—politically and personally. But just as the Clintons started to turn their heads to the future, Pope John Paul II intervened to highlight an old issue: the disconnect between their faith in God and their views on abortion. The Clintons met with Pope John Paul II each time he visited America during their presidency, with the most recent occasion coming during an October 1995 papal appearance in Newark, New Jersey. On each visit, the pontiff raised the abortion issue publicly and privately.[64]

The final meeting of significance between the three came on January 26, 1999, in St. Louis. Once again, Hillary was there. She and her husband joined about five hundred families at a sweltering hot Air National Guard hangar at Lambert Airport to welcome the pope, who had just completed a five-day visit to Mexico. He was scheduled to celebrate Mass before a congregation of more than one hundred thousand at the Trans World Dome. It was his fifth trip to the U.S. mainland; judging by his appearance, which was increasingly frail from the ravages of Parkinson's disease, it appeared that this could be one of the last for the seventy-eight-year-old pontiff.[65]

The weakened state of the pope did not prevent him from following the aggressive agenda on abortion that he had pursued in all their previous meetings. What happened was a near exact repeat of that first encounter between the pope and the Clintons at the airport in Denver in 1994. Again, Bill provided opening remarks. He spoke of human rights and human dignity, though, like his wife, separating them from the abortion issue. "We honor you for standing for human dignity and human rights," said Clinton. "People still need to hear your message that all are God's children."

Again, the pope knew that the Clintons did not apply those words to the same group of human beings that the pope had in mind. And here, too, once again, the pope responded with pointed words aimed at the abortion issue, just as he had done in 1994 in Denver, and as Mother Teresa had done at the prayer breakfast. With the president and Mrs. Clinton listening, the pope—this time his arm shaking and body slightly hunched—made an analogy to the Dred Scott case, which had been tried there in St. Louis. The pope noted that the Supreme Court of this great country of America had, in 1857, declared "an entire class of human beings—people of African descent—outside the boundaries of the national community and the Constitution's protection." Hillary and Bill knew what was coming next: "Today," the Holy Father continued, "the conflict is between a culture that affirms, cherishes, and celebrates the gift of life, and a culture that seeks to declare entire groups of human beings—the unborn, the terminally ill, the handicapped, and others considered 'unuseful'—to be outside the boundaries of legal protection."

The analogy was effective as it connected abortion to one of the issues closest to the Clintons' hearts—racial bigotry. The pope saw abortion as the slavery of the twentieth century, yet his homily probably mattered little. Intellectually, religiously, morally, Hillary had long ago disconnected the abortion question from civil rights, human rights, and her religious worldview. Nonetheless, the speech served as a reminder to the first couple of his admonition five years earlier in Denver: that it remained a "time of testing" for the United States of America.

Afterward, the president and the pontiff held a twenty-minute private meeting during which the abortion issue again "came up," according to White House spokesman P. J. Crowley.[66] John Paul II told Bill Clinton that he hoped that "the value of the human being will be defended and protected in all circumstances."[67] Clinton steered the discussion to the subject of international affairs, namely Cuba and Iraq, commenting later through his spokesman on the pope's profound "moral guidance" on issues of "human justice." Despite these platitudes, there was clearly a rift between the pope and the Clintons that could never be resolved. In the pope's mind, the Clintons didn't get it; they refused to connect the dots that he believed constituted the human fabric knitted by God in the womb.

While Monica Lewinsky and the events of the previous year may have forced Bill and Hillary to reevaluate many of their deepest held beliefs about marriage and each other, they still were not ready to recalibrate their view on abortion. Even with all the religious counseling and spiritual outreach that the pair engaged in during their difficult times, the faith that had pushed them past the sins of adultery could not bring them to go against the values of the Democratic Party. No matter how incongruous their pro-choice spirituality was with the larger teachings of Christianity, the pair would not budge, and as they moved ahead and put the past behind them, this difficult tension between spirituality and politics would continue to raise problems during the years and months ahead.

CHAPTER 12

Transition

As 1999 picked up steam, Hillary began to look past the presidency to the future and plan how she could use the notoriety she had gained as first lady to achieve an elected office of her own. As she carefully weighed her options, opportunity presented itself in the form of a Senate seat opening in New York; Hillary had never been a resident of the state, but neither had Bobby Kennedy, who three decades earlier had looked to the same seat as a means to the presidency. Even in a state as liberal as New York, the seat was not a slam dunk for Hillary. She would need to convince New Yorkers as a whole that they could rely on her, an outsider, for effective representation.

On paper, it did not seem that the overwhelmingly Democratic New York City would pose a problem, but outside the city, things grew more complicated—especially in rural Republican areas, where she was viewed as a liberal carpetbagger. In many parts of the state, her unlikability factor was quite high, and most conservatives viewed her contemptuously. For the Hillary campaign, the key would be to win over just enough undecided voters, along with a massive onslaught of city dwellers.

Thus, 1999 and 2000 were about to become two of the most significant years in Hillary's political life, as she tried to separate herself from her husband's humiliations and her own past baggage by preparing for a run for the venerable U.S. Senate. During this time of new horizons and ideas for Hillary, her faith would be scrutinized as never before, with voters and the media examining it not only to understand her core values, but also to measure her opinions on hot-button religious-political issues such as abortion, gay rights, and gay marriage. In a state with so many ardent Democrats, she would find some of her positions damaging, as her close relationship with God and her stance on the morality of marriage were not in line with some of her party's faithful. But despite public scrutiny of her positions, she would persevere, pushing her agenda straight through Election Day and possibly toward victory.

Gay Rights

Just as her candidacy was beginning to gear up in December 1999, some of her religious values and political stances unexpectedly put her at odds with some of New York's secular Democrats.

At a fund-raiser in New York City's SoHo neighborhood, Hillary told a group of gay contributors that the "don't ask, don't tell" policy, enacted by her husband with the intent of making it easier for gay men and lesbians to serve in the armed forces, had been a failure. In her first public statement on the issue, the Senate candidate said that if elected, she would work to overturn the policy, insisting that homosexuals be allowed to serve openly in the military. Stating that it was politically unrealistic to expect Congress to make a change at the current moment, the first lady maintained that the Department of Defense should take immediate steps to reduce the number of instances of homosexuals being discharged from the military. "Gays and lesbians already serve with distinction in our nation's armed forces and should not face discrimination," said Mrs. Clinton in the statement. "Fitness to serve should be based on an individual's conduct, not their sexual orientation."[1]

While this opinion would not have threatened her reputation with the crowd, her next set of remarks did. Addressing another hot-button issue of gay rights at the fund-raiser, she said that she supported "domestic-partnership measures" that would permit homosexual partners to receive the same health and financial benefits as married couples.[2] However, her spokesman, Howard Wolfson, added that Hillary, like her husband, supported the Republican Congress's 1996 Defense of Marriage Act, banning federal recognition of gay marriage and allowing states to ignore same-sex unions licensed elsewhere. In her speech, she did not use the word "marriage," saying only that "same-sex unions" should be recognized and entitled to all the rights and privileges received by heterosexual couples—though not the right of marriage, as her spokesman noted after the speech.[3]

This was an important distinction and one that would be crucial in the campaign ahead. Gay marriage in the American political and religious debate was quickly emerging as one of the defining social issues of the campaign and beyond. Despite Hillary's advocacy for gay rights, her refusal to support gay marriage placed her squarely in opposition to many in her party. In truth, her unwillingness to support same-sex marriage was consistent with a compelling quote by Don Jones that Gail Sheehy reported in her book released around the time, a remark that did not sit well with gay-rights advocacy groups: "Surely, she is for gay rights, there's no question about that," explained Jones. "But I think both she and Bill still think of heterosexuality as normative."[4]

Her SoHo talk was honest but not as frank as it could have been, prompting the New York press to demand clarification. They got it, as Hillary proceeded to state categorically that she opposed legalizing same-sex marriages: Speaking in White Plains, New York, she provided a strong, clear explanation, that to this day is the most quoted statement enunciating her position. "Marriage has historic, religious and moral content that goes back to the beginning of time, and I think a marriage is as a marriage has always been, between a man and a woman," said the first lady. "But I also believe that people in committed gay marriages, as they believe them to be, should be given rights under the law that recognize and respect their relationship."[5]

Joel Siegel of the New York *Daily News* said that Hillary then "reluctantly waded" into the myriad of follow-up questions. Among them, significantly, she said that she backed her husband's signing of the Defense of Marriage Act. She said what everyone wanted to know: Yes, if she had been in the Senate in 1996, she would have supported the law.[6]

From there, the gay marriage issue heated up even more at the March 2000 annual St. Patrick's Day parade, where organizers barred gays from marching as a single, organized group. As a result, many advocacy groups for gay and lesbian rights boycotted the parade, forcing New York's politicians to decide whether to sit out or join the parade. Caught between these two political constituencies that she needed badly in November, Mrs. Clinton had a decision to make. She decided to march. She was not a popular addition: As the AP reported, Mayor Rudolph Giuliani, her Republican opponent in the Senate race, "basked in the adoration" of the crowd, while Clinton plowed along forty-two blocks up Fifth Avenue to taunts of "Go back to Arkansas!"[7]

In her remarks to the press afterward, Hillary only made it worse, trying to straddle the fence while implying that the Irish organizers were narrow-minded. She said she had "hoped the parade would be inclusive," which she felt it was not, but had opted to march because "this is a day also to honor the values and contributions of Irish-Americans. . . . When you're in public life, you have to balance competing values."[8] At the other end, her compromise angered homosexuals.

Christine Quinn, the City Council speaker, a councilwoman, and a gay activist, told the New York *Daily News* that she was "very disappointed" in Hillary, as were her gay supporters.[9]

It was not a good day for Hillary; she had alienated both sides she had sought to appease. It was no great surprise when in 2001 she avoided the decision altogether, saying she had a scheduling conflict that prevented her from marching.[10]

The Abortion Issue

As Hillary campaigned for herself for the first time, she became even more politically attached to the pro-choice lobby, which she now reached out to more than ever before. Now her thoughts on the abortion issue mattered more than ever, as it was she, and not just her husband, who held the potential to make federal law.

On January 22, 1999, Hillary took an unprecedented step for a first lady by delivering a speech to NARAL, the National Abortion Rights Action League, the premier advocacy group for legal, unrestricted abortion. Speaking to the group in Washington, D.C., she stated her goal of "keeping abortion safe, legal and rare into the next century," a slogan that would become the mantra for her position. It was not a lengthy speech, and did not feature the stridency of the talk she would give to NARAL five years later. Nonetheless, she shared revealing remarks beyond conventional pro-choice sentiments, including the noteworthy idea that she had "met thousands and thousands of pro-choice men and women. I have never met anyone who is pro-abortion. Being pro-choice is not being pro-abortion. Being pro-choice is trusting the individual to make the right decision for herself and her family, and not entrusting that decision to anyone wearing the authority of government in any regard." The clarification was an important one that displayed her distinct aptitude for political foresight. This notion that she was pro-abortion would be precisely the argument that moderate to conservative Christians would use against her in their literature.

Also, her words were helpful in illuminating the contours of her underlying personal philosophy on the abortion question. From her remarks and others like them, Hillary demonstrated her belief that the mother of a household has exclusive authority to decide whether the rest of her family will meet the son, daughter, brother, or sister she is carrying in her womb. Hillary feels that the morally decisive factor in that decision is the *right* to make the decision; the exercising of the right determines the rightness of the decision, regardless of which decision the mother chooses. What is moral is what is decided by each and every mother in each and every pregnancy situation; what is moral is relative to the individual—to each and every individual mother-to-be. The father cannot be the moral arbiter in this decision because he is not endowed with the right of "choice."

There is a personal absolutism here, anchored in moral relativism. The right of reproductive choice is judged the highest moral end in the equation, trumping all other rights, including any consideration of a right to life for the unborn child. This creates a kind of an abortion theology in which the woman having the abortion has the right to decide what is moral for her situation. What is truly right must be judged on a case-by-case basis; there is no objective truth.

Perhaps all of this is a bit too philosophical for one's tastes, but these are important, necessary considerations in trying to understand the interwoven relationship between Hillary's faith and her pro-choice stance. It is precisely this concept of "the mother as moral arbiter" that may help pro-life Christians to understand how Hillary can completely surrender to the pro-choice position, while also submitting to Jesus Christ and to a God that preaches, "Thou shall not kill." Despite discerning this rationale, many pro-life Christians find themselves unmoved by her logic. Pro-life Christians remain frustrated by what they perceive as her selective social justice.

In the NARAL speech, Hillary continued beyond her argument

for being pro-choice, saying, "Fewer teens are having sex, getting pregnant, and having abortions, but there are clearly too many young people who have not gotten the message." She added that young men must also be reached, not leaving males out of the equation: "More has to be done to reach out to young men, and enlist them in the campaign to make abortions rare, and to make it possible for them to define their lives in terms other than what they imagine sexual prowess and fatherhood being."

The sum total was that she had unequivocally laid out her position on abortion rights, demonstrating not only her beliefs but the morality of those beliefs and how she reconciled them to her faith. In the coming months and years, she would continue to revisit and elaborate on these ideas, picking and choosing the elements to emphasize depending on her audience and the issues of the day.

Abortion and the Opposition

While Hillary's pro-choice stance was a big part of her platform, it was by no means unique to her. Her opponent in the race was the Republican mayor of New York City, Rudolph Giuliani, who despite his party affiliation and his Catholicism was also pro-choice. At times during the race, Giuliani squared off with Mrs. Clinton over who was a greater champion of abortion rights, with each pushing the other on the issue. On January 22, 2000, the *New York Times* tried to keep score in a feature on Mrs. Clinton and abortion, titled "Hillary Clinton Vows to Fight to Preserve Abortion Rights." The piece stated, "Signaling that she will not yield the issue of abortion rights in her race for United States Senate, Hillary Rodham Clinton yesterday presented herself as a stronger advocate on the issue than Mayor Rudolph W. Giuliani. Mrs. Clinton said she would make protecting abortion rights a central concern if she is elected to office."[11]

Hillary spoke up in a press conference: "Depending on what happens in the presidential election, what happens on the Supreme Court and in the federal judiciary, we could be facing some very serious challenges in the next couple of years to Roe v. Wade," she said. "I want New Yorkers to know that I wouldn't only vote right, but I would be a strong voice, and I would attempt to organize as much as I could to be sure that we defended a woman's right to choose." She added: "I want New Yorkers to know that I am and always have been pro-choice, and that it is not a right that any of us should take for granted. There are a number of forces at work in our society that would try to turn back the clock and undermine a woman's right to choose, and we must remain vigilant."[12]

She again affirmed definitively: "I have always been pro-choice and my position has never changed, and never wavered." She was emphatic, clearer than on any other subject, repeating: "The current membership of the United States Senate is not something that we can count on to protect a woman's right to choose. For women, the Senate will be our court of last resort, both because of the votes that will be taken there on issues, and because of the votes that are likely to be taken on judicial nominees."[13]

In May 2000, shortly before the primary, the political situation for Republicans hit a snag when a diagnosis of prostate cancer forced Giuliani out of the race. The party replaced Giuliani with Representative Rick Lazio, a candidate who was also pro-choice. Maneuvering to win the issue, Hillary soon moved to prove to the state that she was more pro-choice than Lazio. It did not take long for Hillary to highlight some of the crucial distinctions. Lazio opposed using federal Medicaid funds to pay for poor women's abortions and favored a ban on partial-birth abortion. Clinton differed on both. Also, she made clear that she would never vote to confirm an anti-abortion nominee to the Supreme Court, while Lazio said there should be no such litmus test for possible appointees.[14]

With Hillary in the running, the pro-choice lobby would always be represented.

Charges of Anti-Semitism

As she sought to win the heart of the state, Hillary's campaign was momentarily derailed by Paul Fray, a man who had managed Bill's first but unsuccessful campaign for Congress in 1974, who now stepped up to confirm rumors that Hillary had once called him an "f–ing Jew bastard." The incident almost died down under Hillary's strong denial, as well as the involvement of leading New York Democrats like Chuck Schumer. "It did not happen," she said of Fray's allegation. "I have never said anything like that. Ever, ever."

Bill also denied it, but with a curious reassurance: "She might have called him a bastard. She's never claimed that she was pure on profanity. But I've never heard her tell a joke with an ethnic connotation. . . . She's so straight on this, she squeaks."[15]

Just as the controversy was beginning to heat up, Arkansas state trooper Larry Patterson weighed in on the allegation, stating that he was hardly surprised by Fray's charge, since he had often heard Hillary and Bill call each other "Jew bastard" and "Jew motherf——." Patterson said this was common parlance for Arkansas's first lady, from whom he said he heard anti-Semitic slurs "at least twenty times."[16]

While the story was carried widely in newspapers like the *New York Post* and on talk radio, where it became known as the "FJB" incident, it failed to garner mention in outlets such as the *New York Times, Newsday*, the Associated Press, and the national TV networks. The campaign's immediate response was to use the incident to her advantage, portraying her as a victim of dirty campaign tactics by Republicans, but the reality was that the GOP had nothing to do with it. Fray was a Democrat who had campaigned for Bill Clinton and voted for him twice.[17] When this initial argument failed to gain traction, Hillary's campaign tried an alternative tactic: One of her advisers, Karen Adler, reportedly circulated a memo to certain Jewish supporters of Hillary, asking them to telephone people in the media and to reach out to the magazines *Forward* and *Jewish Week* in particular, for the purpose of telling them that Hillary loved and admired Jewish people. According to published reports, the memo instructed these supporters to conceal from journalists the fact that the campaign had asked them to make these calls.[18]

Shortly after the FJB story exploded, Hillary made one of the first of many trips to a house of worship during the Senate campaign. She ventured to the Hampton Synagogue in Westhampton Beach, a congregation run by Marc Schneier, a prominent liberal rabbi. Predictably, she did not risk trying a hard-core Orthodox synagogue.[19] Important to the bigger picture was the positive effect of Mrs. Clinton's trip to this synagogue: She and her staff were sure that this outreach had healed the damage. It was a successful political move, and it was just the beginning of her religious campaigning, which took her literally right inside houses of worship.

Church Campaigns

As Election Day approached, Hillary began working churches like a preacher, employing her faith for political purposes in ways she had never done before. She did so with no objection from the intensely secular, religiously hostile New York press.

No observer captured this as well as Beth Harpaz, who wrote the 2001 book *The Girls in the Van: Covering Hillary*. Harpaz recorded in her journal one day:

October 1, 2000. It's a Sunday morning and I'm with Hillary. And that means I'm in a black church, because that's where Hillary goes every Sunday between Labor Day and Election Day. Some days we start with the 7am service and hit our sixth or seventh church around mid-afternoon, but today the schedule is light: just three churches before noon.[20]

Harpaz goes on to describe an unparalleled routine of church visiting during Hillary's 2000 Senate run, stating that Hillary was "not content to hit a half dozen churches" like a typical Democratic candidate. In the two months leading up to Election Day, the first lady visited more than twenty-seven African-American churches, from the smallest storefront tabernacles to the most prominent parishes in the city.[21]

Clearly this was an intense and focused strategy to pick off votes at Sunday services, and yet this behavior elicits a curious question: Why does the mainstream press seem to have a problem with Republicans who simply talk about their faith, while religious Democrats like Hillary remain unscathed when they go so far as to vigorously campaign in churches? Perhaps the press was unaware of Hillary's church politics? Not at all. As Harpaz matter-of-factly noted in her next sentence: "As always, the row of seats taken up by the press corps is just about the only part of the church occupied by white faces." Besides, she added, "Democratic politicians usually visit" churches; indeed, this is the best-kept secret in the press corps.[22]

Not only did the media not object to Hillary's annihilation of any barriers separating church and state, but they loved it, and sometimes even cheered her on. Consider Harpaz's report on the October 1 Hillary gathering at Memorial Baptist Church in Harlem: "Most of the worshipers are on their feet, clapping and singing, and rocking to an electric guitar, piano, and drum ensemble driven by the steady, happy jangle of a tambourine. 'Lift Him up!' the several hundred voices sing as one, and within minutes, I and most of the other reporters stand up, too, clapping and swaying along with them, the music is irresistible."[23]

Blame the music, but it is hard to imagine the press reacting quite the same way to, say, an appearance by John Ashcroft at the campus chapel at Jerry Falwell's Liberty University, or a "church tour" by a Republican campaigning through white Baptist churches in the South. The fact is that the appearance in itself would not be problematic if the press's response were not so laden with hypocrisy. Campaigning in churches was shrewd politics for Hillary, but she was largely enabled by a press corps that she knew would not portray her as violating the church and state barrier that is so important to many mainstream Democrats. Safe and relatively confident that the appearances would not result in a church vs. state debate, Hillary made churches a focal point of her campaign strategy. Events like the one in Harlem helped Mrs. Clinton, and the journalists sat quietly, approvingly, either not even realizing or at least ignoring the stunning double standard they applied to politicians they liked and disliked.

Perhaps Hillary's talks in these churches somehow managed to carefully avoid politics? Again, not at all. Politics was the stated purpose.

Harpaz continued her description of the moment, as Hillary appeared behind the podium:

> "She's gonna win," declares the pastor. "And we are going to come out in droves for her."
>
> It's a point that needs to be made. Nobody is doubting that black voters prefer Hillary over Lazio. But black turnout is unreliable in New York City. . . .
>
> "Whooo!" Hillary hoots as the applause dies down and she looks around, feeling the love. "Thank you for the day the Lord has made!"
>
> It's her standard opening line, a riff on the psalm. . . . Now we are about to hear the press corps' favorite part of Hillary's Standard Sunday Morning Sermon.

Here Harpaz elaborated on Hillary's stump speech that she used in almost every African-American church that she visited, a speech tailored to African-American voters that touched on slavery, Harriet Tubman, freedom for nineteenth-century blacks. During the speech, she would recall her trip to Auburn, New York, where she went to see the house where Harriet Tubman lived after she escaped from slavery. She described Tubman's bravery, and her willingness to sacrifice her own life for the greater good of other slaves. It was a riveting speech for many reasons, but especially because of the clarity with which she tied this historical narrative to herself and the present day. According to Harpaz, Hillary went on, "We all have to keep going until we are a just nation," practically shouting as people began to stand, cheering and clapping and loving every word out of her mouth. "But I need your help. And if you will help me, I will be there for you!" The church exploded into applause at Hillary's closing exhortation: "Let's keep going! Let's have a great big turnout in the election! If you fight for me in the next five weeks, I will go to the Senate and fight for you for the next six years! Thank you, everybody! Thank you *so-o-o-o-o much!*"[24]

This was the finale, as Hillary would always end with a cry of support and the roar of the band before climbing into her van with the energized secular media straggling behind her, trying to keep up en route to the next service.

This church stop template that Harpaz described changed depending on location. For example, at the next Sunday worship-service/rally that same October 1, a more staid Episcopal church, where the bishop told the gathered they were about to experience a "visitation" from Hillary Rodham Clinton, Hillary was shrewd enough to save the "Keep going!" routine for churches like the previous one. Here, said Harpaz, "Instead, she trots out another one of my favorite Hillary church-shticks, which goes something like this: 'Someone asked me the other day if I prayed. I said, yes, I do pray. I was fortunate enough to be brought up in a home where the power of prayers was understood. But I have to tell you, if I hadn't prayed before I got to the White House, I would have started after I arrived.' It's a funny line, and even in this buttoned-up place, people laugh."[25]

At times, these events got out of hand, with statements from liberal pastors that were extremely mean-spirited. One such occurrence that Harpaz reported was during a stop at Emmanuel Baptist Church in New York City, where co-pastor Darlene Thomas McGuire—after issuing the standard claim that she was not speaking for the church, judged Hillary's opponent evil. Actually, it was more than that: In a hymn, McGuire directly substituted Hillary's opponent, Rick Lazio, for no less than the Prince of Darkness himself. McGuire, immediately after claiming, "I'm not speaking for the church today," led Hillary and her entire congregation in a unique rendition of an old-time hymn:

> I told Satan, get thee behind
> I told Satan, get thee behind
> Get thee behind
> Get thee behind
> Victory today is mine!

McGuire then led her flock in a revised second verse, belting out the new lyrics loudly and proudly, in a state of near political-religious ecstasy:

> I told Lazio, get thee behind!
> I told Lazio, get thee behind!
> Get thee behind!
> Get thee behind!
> Victory today is mine![26]

In a question-and-answer session after the appearance, Harpaz asked Hillary what she thought of the new lyrics. "She paused for a second," said Harpaz, "then smiled and replied, 'I love hymns.'"[27] So did the press, which, uplifted by that old-time religion, was experiencing an old-fashioned conversion; none of the other reporters voiced any questions about comparing Lazio to Satan.

At these church rallies, minister after minister—men and women of God—made the questionable claim that Hillary's campaign stops at their churches should not be construed as a political endorsement. Nevertheless they called her everything from "a woman of God," to, as one reverend said at the Metropolitan AME Church in Harlem, "another Joshua." When Michael Kelly had dubbed Mrs. Clinton "Saint Hillary" seven years earlier, he had merely been a bit premature.

Like Joshua being chosen by God to lead Jews after Moses' death, Hillary said she would lead this group of African Americans to the Promised Land—directly implying that God had chosen her for that mission. However, she warned them, it was up to them; they could not sit on their hands. It was not enough, she told them, to pray to God to help them. No, she said, in a line she saw her husband use in church after church, "We also have to move our feet and hands." This was a not-so-veiled way of telling them that they needed to walk to the polling booth and pull that lever for her and other Democrats.[28]

When she finished, the pastor—again, claiming to be nonpolitical—conceded that he agreed with Mrs. Clinton. Wrote Beth Harpaz, again a witness: "When she sat down, the pastor, Robert Bailey, got up again and told his flock that God had given them a second chance to elect another Clinton. Hillary, he said, was their Joshua."[29]

A typical example of how the press reacted to all of this was the *New York Times* coverage of a Hillary church rally on Sunday, November 5, the last Sunday before the election. As Texas governor George W. Bush made routine campaign stops in Florida—deliberately avoiding church stops, no doubt fully understanding he could never get away with a similar campaign strategy—on that day, senatorial candidate Clinton appeared at as many New York churches as she could. The *Times's* coverage began: "In a day of gospel and politics, Hillary Rodham Clinton preached and prayed her way through seven churches in seven hours yesterday, moving to close out her campaign by urging black parishioners in New York City to turn out to support her tomorrow." From Brooklyn to the Bronx, said the *Times*, at churches large and small, the first lady "pleaded and cajoled church-goers to vote the Democratic line. . . . The same striking scene was repeated again and again throughout the day."[30]

The *Times* quoted Mrs. Clinton herself: "It would be a shame if we stayed at home this Tuesday," she shouted into a microphone from the pulpit of the massive Allen AME Church in Queens. As she finished her political sermon, the reporter watched her grin, set down the microphone, and shrug, "One more church, one more rally." The reporters smiled as well, as did the beaming Reverend Charles E. Betts Sr., who raised his hands to proclaim to the heavens, "God is raising up another woman of God."[31]

New York Times reporter Adam Nagourney recorded the "rustle of excitement and raised hands and swell of organ music and gospel song" that accompanied Mrs. Clinton at the worship services. The *Times* even ran a photograph of an African-American woman from a Bronx church holding a sign that read, "All Souls to the Polls."

Not a single *New York Times* editorial or columnist expressed outrage at the "misuse" of religion. Given that many in the media were used to similar appearances by Bill Clinton, their inaction was hardly surprising. Hillary had learned the craft at the side of the master, and trusted the press to keep silent on this politicization of faith, so long as the tactic helped the Clintons and Democrats get elected.

The Other Clinton

As Rick Lazio closed within single digits of Hillary's lead, she narrowed her focus even more intently on the African-American vote.[32] In 1992 and 1996, her husband had received 83 percent and 84 percent, respectively, of the black vote. In the last presidential election, 1996, Bill Clinton actually lost white Americans 46 percent to 44 percent, making his biggest gains among the black community, which he won 84 percent to 12 percent.[33]

Thus, Mrs. Clinton and the 2000 Democratic presidential nominee, Vice President Al Gore, needed similar numbers for 2000. The Gore campaign, which trailed George W. Bush in all but one major poll the day before the election and nearly all of the previous two weeks, decided that they needed an unprecedented mobilization of black voters to give their guy a chance at the overall popular vote, a brilliant analysis that turned out to be absolutely accurate, and was the unforeseen reason in Gore securing the popular vote in 2000, contrary to the prior expectations of nearly all pollsters.[34]

To mobilize that black vote, Bill did his part, appearing in African American churches throughout Hillary's new territory. For instance, he hit a black church in New York City on October 31, 2000, the Kelly Temple Church of God in Christ in Harlem. As in most such talks, he was joined by a contingent of fellow Democratic politicians. He began by reminding congregants why they were there:

> Now, we all know why we're here, and we can shout amen and
> have a great time, and we're all preaching to the saved. . . . But
> I want to talk to you about the people that aren't in this church
> tonight, the people who have never come to an event like this and
> never heard a President speak or even a mayor or a comptroller
> or a Senator or anybody. But they could vote. And they need
> to vote, and they need to know why they're voting. And that's
> really why you're here, because of all the people who aren't here.
> Isn't that right? . . . So what you have to think about tonight
> is, what is it you intend to do between now and Tuesday, and
> on Tuesday, to get as many people there as possible and to make
> sure when they get to the polls, they know why they're there,
> what the stakes are, and what the consequences are. . . . If you've
> got any friends across the river in New Jersey or anyplace else, I
> want you to reach them between now and Tuesday, because this
> is a razor-thin election.[35]

As was common for Bill, he made appearances such as this throughout the campaign season. Just two days earlier, on October 29, the president campaigned in two churches at two different Sunday services, on these occasions for the purpose of rallying votes for Vice President Gore. Speaking to the congregation of Alfred Baptist Church in Alexandria, Virginia, Clinton employed a Bible verse as justification to head to the polls: "The Scripture says, 'While we have time, let us do good unto all men.' And a week from Tuesday, it will be time for us to vote."[36]

Clinton was joined at the Alexandria church by a prominent collection of Democratic politicians, including U.S. Representative Jim Moran (D-Va.), who was up for reelection. That talk came at 12:40 P.M. Earlier, at 9:40 A.M., he squeezed in another campaign talk to the congregation of the Shiloh Baptist Church in Washington, D.C., which included, as he openly admitted, "so many members of the White House staff," apparently too many for him to get an accurate snap count, as well as the D.C. delegate to Congress, Eleanor Holmes Norton. Clinton gave a pitch for various types of federal legislation, including the D.C. College Access Act, and blasted Republican-proposed tax cuts, before closing by urging the worshippers to get out the vote on Election Day.[37]

The speeches in these churches had little, if anything, to do with religion. When Scripture was mentioned, it was usually for strictly political purposes.[38] And, to that end, as he told a congregation in Newark, Bill Clinton believed that he and fellow Democrats were doing the Lord's work: "God's work must be our own." That, said the president, was a central motivating factor in both the "last presidential election of the 20th century, and the first presidential election of the 21st century"—in other words, in choosing Bill Clinton over Bob Dole in 1996 and ultimately Al Gore over George W. Bush in 2000.[39]

Overall, Bill Clinton spoke in churches twenty-one times as president, more than half of which occurrences (twelve) came in election years.[40] (By comparison, the *Presidential Documents* list three incidences of President George W. Bush speaking in churches through his first three years in office—none for the purpose of campaigning.[41]) Hillary quickly surpassed that: She did twenty-seven churches in just two months, and had not even been elected yet.

Along with Vice President Al Gore, who also barnstormed churches, the Clintons helped form a kind of a political trinity for the Democrats in 2000, the party's three standard-bearers, its three top political figures, discipling to the faithful, offering political salvation inside the churches of their most devout constituency. They ministered to a voting bloc that gave them almost total political devotion, a group that gave them their vote with a religious reliability, unquestioning the three's moral soundness.

To be fair, Hillary and Bill were much more cautious with their words and appearances than Vice President Gore, whose campaign talks in chapels from Pittsburgh to Philadelphia and Detroit to Memphis seemed more like scare tactics designed to frighten African-American worshippers into thinking a George W. Bush presidency might bring back Jim Crow laws and possibly even lynchings.[42] In Philadelphia two days before the election, Gore appeared with the sister of James Byrd, a black man who had been dragged to his death by racists in Texas, where Bush was governor. Gore described in vivid detail Byrd's mutilation, a graphic moment described by the *New York Times* as "the emotional high point" of Gore's day. Why was Gore reminding African Americans of this horrible incident? The *Times* explained: The Texas murder during Bush's governorship was how Gore "rallied his base."[43]

In addressing the congregation in Memphis, Gore boiled down the choice between him and George W. Bush as one between good and evil. "Deep within us," said Gore, "we each have the capacity for good and evil. I am taught that good overcomes evil if we choose that outcome. I feel it coming. I feel a message from this gathering that on Tuesday we're going to carry Tennessee and Memphis is going to lead the way." It was up to the worshippers to ensure that evil—a Bush victory—would not prevail.[44] Befitting one of the legs of the political trinity, Gore possessed the ability to separate the wheat from the chaff, to see into a man's soul, and to judge and recognize good and evil.

These resonant images seemed to be working beautifully for Gore. After hearing the vice president speak at their church in Pittsburgh, twenty-three-year-old Raushana Ellison of Pittsburgh's Hill District and sixty-seven-year-old Willa Mae Tot of the Overbrook section of Pittsburgh said they were scared.[45] They would be voting for Vice President Gore.

To her credit, Hillary Rodham Clinton did not descend as low as Gore, even as some of the ministers who hosted her went way overboard. Such depths were out of her range in the 2000 campaign.

It is also worth noting that Hillary is consistent and not hypocritical on the subject of church appearances: She has never stated that politicians, Democrat or Republican, cannot or should not campaign in churches. During this campaign, she used her faith in a way that many Republicans never have the opportunity to, and she used this ability to its fullest. Because she had the freedom to campaign in the churches themselves, she was able to proclaim her faith directly to the people who would care about it most. Instead of seeking out alternate venues that are sympathetic to religious discussion, Hillary was able to go straight to the source to demonstrate the extent of her piety. It gave her a decided advantage, and it seemed undeniable that the media made this possible.

No matter how one feels about politicians campaigning in churches, this much is certain: The strategy—risk-free, thanks to a passive press—was working wonders for Mrs. Clinton's first bid for elected office, and thus one could expect more in campaigns ahead.

CHAPTER 13

Senator Clinton

On November 7, 2000, First Lady Hillary Rodham Clinton became Senator-elect Hillary Rodham Clinton (D-N.Y.), whipping Republican challenger Congressman Rick Lazio by a vote of 55.3 percent to 43.0 percent, winning the ballots of 3,747,310 New Yorkers.[1] As for her husband, he was term-limited out of the presidency, and his vice president, Al Gore, the third leg of that triumvirate, lost his bid to succeed Bill Clinton in one of the closest, most divisive presidential contests in history. Gore was bitter over the defeat, having lost the electoral college but not the popular vote. Bill Clinton, however, was very gracious, as was evident during the January 20, 2001, inaugural, where the outgoing president sat next to his wife, the incoming senator for the state of New York.

The next time the three were seen together by so many eyes came on September 14, 2001, three days after one of the most dreadful days in American history. Hillary was obviously outraged by the attacks and condemned them and their perpetrators. But because she allegedly feared being heckled, as she was such a polarizing figure and had not been received warmly (actually quite harshly) by New York City police and firefighters, she reportedly did not attend any of the hundreds of funerals stretching out over the next several months for the nearly three thousand victims, largely remaining in Washington, according to biographer Christopher Andersen in his highly critical book on Hillary. To the contrary, the other Democratic senator from New York, Chuck Schumer, showed up at a dozen funerals, and Governor George Pataki and Mayor Rudy Giuliani paid their respects more times than anyone could count.[2]

For someone who had recently found herself in churches multiple times per day during the political season, this was an about-face that did not go unnoticed. One service that she did attend, however, was the official memorial at the majestic National Cathedral, held September 14, 2001, a day that President George W. Bush had declared a National Day of Prayer and Remembrance. Bush personally organized the service, even picking the music and speakers, including a multicultural gathering that featured a woman bishop, two black ministers, a rabbi, a Catholic bishop, and a Muslim imam. This inclusiveness was deliberate, particularly the inclusion of Imam Muzammil Siddiqi of the Islamic Center of North America.[3] The seventh and final religious figure to speak, before Bush rose, was an eighty-two-year-old Billy Graham.

Bush ensured that the Clintons sat in the front pew. Typical of the two, Hillary was unemotional, almost expressionless, whereas Bill was just the opposite. Throughout the service, he wore his heart on his sleeve, tears at times streaming down his cheeks during hymns. Breaking protocol, his emotions got the best of him as he jumped up and led a standing ovation when the aged Billy Graham finished talking. It was classic Bill.

Shortly after September 11, Hillary supported President Bush's use of military force in Afghanistan in subsequent weeks as well as his pursuit of Osama Bin Laden and al-Qaeda. About a year later, she would vote to authorize President Bush to wage war in Iraq.

Throughout her first term in the Senate, she worked hard to separate herself from the party's hard left on Iraq and the War on Terror. She visited troops in Iraq and Afghanistan, proposed pay increases for soldiers and better health benefits for the National Guard, and forcefully opposed a number of base closings.

To her credit, she has not inappropriately inserted Jesus into the war debate, as have many figures on the religious left, such as the group Religious Leaders for Sensible Priorities, which in December 2002 placed an ad in the *New York Times* that judged, "President Bush: Your war would violate the teachings of Jesus Christ."[4] Hillary seems to have realized that it would be a bit rash to presume to know the Lord's divine thinking on whether U.S. Marines should have pursued Saddam Hussein.

While these decisions and others like them led her to position herself successfully as a moderate on defense issues, it was the economic and social issues that continued to endear her to much of the Democratic Party, even as her religious traditionalism forced her to deviate on the issues of gay marriage, "the culture of violence and sex in the media," and her longtime favoring of abstinence as a means to prevent teen pregnancy. Of course, the vast majority of Americans and members of Congress oppose gay marriage and raunchy movies, and support teen abstinence, meaning that Mrs. Clinton's positions in these sensible areas did not nudge her to the right so much as they took her out of the fringe minority.

Nevertheless, supporters in the press did their best to portray her as a pragmatist seeking the middle ground, a picture that best served her electoral prospects. The Associated Press dubbed her "a Northeastern centrist." Likewise, the *New York Times* proclaimed that Mrs. Clinton "has defied simple ideological labeling since joining the Senate, ending up in the political center on issues like health care, welfare, abortion, morality and values, and national defense, to name a few."[5]

Though the Associated Press and the *New York Times* purport to be nonpartisan, or at least more objective than a conservative newspaper like the *Washington Times*, it was the *Washington Times* that assessed Mrs. Clinton's alleged centrism in a quantifiable, objective manner, drawing from ratings from various groups, left and right, that rank the liberalism-conservatism of politicians. The *Washington Times* compared Mrs. Clinton's rankings to the ultimate noncentrist from the Northeast, Senator Ted Kennedy (D-Mass.). Here is what it noted:

Americans for Democratic Action (ADA), the liberal watchdog that rates members of Congress according to their voting patterns, assigned Hillary a near-perfect 95 percent liberal quotient for each of her first three years in the Senate, close to the perfect 100 percent of Ted Kennedy for 2001 and 2002 and Kennedy's 95 percent for 2003. The American Conservative Union, ADA's counterpart, gave her an 11 percent conservative ranking for the same period, compared to a 5 percent for Kennedy. Another left-leaning group, the AFL-CIO, gave Senator Clinton a 93 percent, precisely the same score as Senator Kennedy. Among conservative organizations, the National Taxpayers Union gave Hillary an average annual rating of 14 percent over her first three years, compared to 13 percent for Kennedy, and the National Tax Limitation Committee gave both a zero for the 107th Congress.[6]

National Journal, a nonpartisan source known for its ratings, found that in 2002, not a single U.S. senator was more liberal on economic and social matters than Mrs. Clinton, and in 2003, no senator surpassed her liberal ranking on social issues. And while *National Journal* ascertained that she was not the most liberal senator on economic matters in 2003 as well, she still ranked very high at 90 percent. Her composite liberal score was 88.8 percent, compared to the scores of two true centrists, Maine Republican senators Olympia Snowe and Susan Collins, who rated as nearly perfectly middle as one can get at 50.5 percent and 50.8 percent, respectively.[7] Only on defense and foreign policy issues has *National Journal* ranked Hillary as more conservative than most of her Democratic colleagues in the Senate.[8]

On the religious/social-conservative front, Clinton and Kennedy received zeros from the Christian Coalition and the National Right to Life Committee, while both received over the same period perfect 100 percent rankings from NARAL.[9]

Another interesting comparison was done by the New York *Daily News*, which compared Mrs. Clinton to Howard Dean, whom *The New Republic* called "one of the most secular candidates to run for president," a former Congregationalist who, as the *Washington Post* put it, "rarely attends church services, unless it is for a political event."[10] As the *Daily News* pointed out, Dean has often been openly scornful of the religiosity of Republicans, dubbing the GOP "a white Christian party," while also judging Republicans "evil." And yet, noted the *Daily News*, Mrs. Clinton has likewise leveled some harsh charges at those on the other side of the aisle, in one instance stating that "some" Republicans "honestly believe they are motivated by the truth," and that "they are motivated by a higher calling, they are motivated by, I guess, a direct line to the heavens."[11] The religious sarcasm seemed to thrill her devout secular base as she struck a tone resonant of Dean, if not quite as harsh; however, statements such as these overlook the fact that she, too, has drawn a direct connection between her own politics and motivation from a higher calling.

Nonetheless, the first-term senator looked to steer toward the middle on some hot-button domestic issues—many of which were linked to a religious component.

The Abortion Votes

With a conservative Republican president replacing Bill Clinton in the Oval Office, pro-life congressional Republicans were invigorated. In 2001, they had been delayed by the events of September 11 and subsequent war in Afghanistan, but by mid-2002, they began launching major bills aimed at curtailing abortion. As they did, the pro-choice lobby knew it had a reliable ally in the senator from New York.

On June 21, 2002, the Senate voted fifty-two to forty to ban abortions in military medical facilities. Mrs. Clinton voted against the ban. On March 12, 2003, in an especially important symbolic gesture, the Senate voted fifty-two to forty-six in favor of the Harkin Amendment endorsing *Roe v. Wade*, which stated, "It is the sense of the Senate that the decision of the Supreme Court in *Roe v. Wade* (410 U.S. 113 (1973)) was appropriate and secures an important constitutional right; and such a decision should not be overturned." The amendment was supported by nine Republicans, forty-two Democrats, and one independent, and was rejected by forty-one Republicans and five Democrats. Mrs. Clinton voted for the amendment.

The next day, on March 13, 2003, came another significant abortion vote, and this one proved a major blow to Mrs. Clinton's ability to position herself as a moderate: She voted against a ban on the grim procedure of partial-birth abortion, which the *New York Times* euphemistically calls "a certain kind of late-term abortions," and the Associated Press (for some reason) has felt the need to qualify as "what critics call 'partial-birth abortion.'"[12] Hillary herself called the procedure "partial-birth abortion" (which is the proper term), and also labeled it "horrible."[13]

The language and categorization of this procedure signals the trouble Mrs. Clinton faces in voting against the ban: Even the legendary New York Democratic senator Daniel Patrick Moynihan, who preceded Hillary, called partial-birth abortion "infanticide." Another major Democrat, former president Jimmy Carter, said of the procedure: "late-term abortions"—"where you kill a baby as it's emerging from its mother's womb."[14] That is precisely what happens, as the baby is partially delivered just far enough that a suction can be inserted into the base of the skull to remove the child's brain before full delivery.

In a debate with Rick Lazio during her 2000 race, Mrs. Clinton had retreated to her husband's position on the issue: "I have said many times that I can support a ban on late-term abortions, including partial-birth abortions," said Hillary, "so long as the health and life of the mother is protected. . . . Of course it's a horrible procedure. No one would argue with that. But if your life is at stake, if your health is at stake, if the potential for having any more children is at stake, this must be a woman's choice."[15]

This was the reason cited by President Clinton for twice vetoing the ban passed by the Senate during his presidency. In response, evangelicals were livid. James Dobson of Focus on the Family stated, "Clinton's hands are stained with the blood of countless innocent babies. By twice vetoing a bill that would have banned partial-birth abortion, he almost single-handedly preserved a barbaric procedure by which fully viable and un-anaesthetized infants, each fresh from the Creator's hand and brimming with life, were murdered during the final moments of delivery."[16] Mrs. Clinton's supporters can expect that Dobson's tough language will likely find reincarnation in the direction of Hillary as she runs for president.

Likewise Mrs. Clinton, her husband, and groups such as NARAL all argued in opposition to the ban, but their wisdom has been disputed in testimony by medical professionals, including the former president of the American Medical Association, Daniel Johnson, who stated in the *New York Times* that the procedure is never medically necessary and is never needed to save the life of the mother nor to ensure her health.[17] Supporters of a ban on the procedure also believed that the "mother's health" exemption being demanded by abortion rights activists would be abused by mothers and abortion doctors to the point that partial-birth abortion could and would always remain a legitimate abortion option, thus rendering the ban ineffective.[18]

Despite the passion of their argument in opposition to the ban, the Clintons have shown themselves to be sensitive to criticism on this issue, especially Bill. A dramatic, unreported illustration took place at a Christmas Eve service at the National Cathedral after Bill's first veto of the partial-birth abortion ban.[19] Reverend Rob Schenck, a leading pro-life activist, was approaching the Communion rail along with a small group of clergy. As Schenck walked past a seated President Clinton, he whispered, "Mr. President, God will hold you to account for the babies." In response, an incensed Clinton ordered the Secret Service to apprehend Schenck and search and interrogate him, which it did. To this day, he remains on a "flag list," meaning that various security details frequently deny him access to Washington events to which he is invited.[20] While Schenck insisted that he addressed Clinton in a nonthreatening way, it could have been Clinton's conscience that felt threatened, perhaps harking back to those heart-to-heart discussions with Reverend W. O. Vaught in the early 1980s.[21]

As for Mrs. Clinton, her conscience has not been affected enough to switch her vote on the issue. In October 2003, Hillary joined a minority of senators in a final vote against the ban on partial-birth abortion, and this time the ban was not vetoed by the president—as the president was George W. Bush, not Bill Clinton.[22]

Because of Mr. and Mrs. Clinton's adamant position that mothers should not be criminalized for abortions, it is worth noting that the partial-birth abortion ban issues fines and up to two years in prison for those who perform the procedure, while stating unequivocally that the women who receive the procedure are not criminally liable.

The Slippery Slope of Marriage Rights

In 2003, another issue tied to religious values came to the forefront of Americans' minds, when Hillary introduced a bill in Congress that would give homosexual couples the same rights as heterosexual couples, a move that some conservatives, including the editorial board of the *Washington Times*, complained would merely provide momentum to efforts designed to legalize gay marriage, an assertion that Mrs. Clinton rejected.[23]

In the *New York Post*, Deborah Orin charged that suddenly a lot of Democrats, starting with Senator Clinton, were "desperately seeking" a "don't ask, don't tell" policy on the subject of gay marriage. Orin said this included Mrs. Clinton's husband, the ex-president. Orin reported that Bill Clinton's office would not say whether he still backed the Defense of Marriage Act he signed into law in 1996. According to Orin's reportage, Hillary herself, who during her Senate campaign had been very clear in her stance on the issue of gay marriage, suddenly was unwilling to weigh in on the Defense of Marriage Act. Orin quoted Hillary spokeswoman Karen Dunn, who said, "This issue is in a state of evolution."[24]

What was going on? That question was pondered by Andrew Sullivan, the former *New Republic* editor and probably the country's most outspoken gay conservative. "[I]t's no surprise to hear her complete non-answer on the question of same-sex marriage," wrote Sullivan in the *New York Sun*.[25] He pointed to the following exchange from a June 18 interview with Mrs. Clinton on the Brian Lehrer WNYC show in New York City:

Lehrer: The lead story in the *New York Times* today is about Canada's decision to fully legalize gay marriage. Do you think the United States should do that?

Clinton: Well, obviously in our system it is unlikely ever to be a national decision because of the way our federal system operates, where states define what the conditions for marriage, or domestic partnership, or civil union might be, so I don't think that we will ever face it. In fact there is a law on the books, passed before I was in Congress, the Defense of Marriage Act, which goes so far to say that even if one state does it, other states under our full-faith and credit clause of the Constitution don't have to recognize it.

Lehrer: But is Canada setting a good example, one that you'd like to see spread through the states here?

Clinton: Well, I have long advocated domestic partnership laws and civil unions, to me. . .

Lehrer: . . . That's different from marriage.

Clinton: Well, marriage means something different. You know, marriage has a meaning that I . . . I think should be kept as it historically has been, but I see no reason whatsoever why people in committed relationships can't have, you know, many of the same rights and the same, you know, respect for their unions that they are seeking and I would like to see that be more accepted than it is.

Lehrer: But not within the context of marriage.

Clinton: Yeah, I, I think that is, you know. . . . First of all, I think that it is unlikely, if not impossible, to be something nationally accepted in our country, but I also think that we can realize the same results for many committed couples by urging that states and localities adopt civil union and domestic partnership laws.[26]

Her words almost suggest (but not necessarily) that if the public changed its attitude on the issue, she would follow suit. On the other hand, her words seem to make clear her continuing commitment to the idea that marriage in America should be legally restricted to a man and a woman.

On this issue, perhaps more than any other, the malleability of Hillary's religious values was apparent. Since 1996, she had acted and spoken in support of the Defense of Marriage Act many times, but as the Democratic Party shifted and gay marriage became an issue that more mainstream members of the party were taking up, Hillary was no longer arguing against it with the determination that once defined her position. Though in the 1990s and during her campaign she had drawn a clear link between her opposition to the issue and her religious and moral definitions of marriage, she found herself hesitating as the political culture that produced that bipartisan bill began to erode. She was no longer arguing in favor of the Defense of Marriage Act with the diligence and consistency she once possessed; on the contrary, she appeared to be leaning toward it. While it was debatable whether the legislation she introduced would have amounted to paving the way for same-sex marriage, it was evident from her ambiguous stance on the issue that her formerly entrenched position might be shifting.

Rather than sticking to the beliefs that she had professed three years prior, she was perhaps beginning to hedge her ideas to coincide with the turning tide of her party.

Keynote Address to NARAL

On January 22, 2004, Mrs. Clinton gave the keynote address at the NARAL dinner celebrating the thirty-first anniversary of *Roe v. Wade*. As the most vitriolic, inflammatory speech she has given on the subject, the speech will no doubt be one of Hillary's larger hurdles in positioning herself as a moderate, religious Democrat. She described pro-lifers—the "anti-choice" people—as plotting behind closed doors, using "foot soldiers" to quietly plan the overthrow of the right to an abortion. This was merely step one, said the senator, in an insidious plot to strip a panorama of rights. She began:

> While the choice debate has changed little since *Roe v. Wade* was decided 31 years ago, the tactics employed by our opponents have changed. They have realized it cannot be done quickly and in the light of day. They can't just propose a constitutional amendment, and make the debate public. No. Our opponents are patient. They are going to do it slowly, quietly, one justice at a time, one legal battle at a time, one state at a time.
>
> As we gather today, forces are aligned to change this country and strip away the rights we enjoy and have come to expect.

Slowly, methodically, quietly, they have begun chipping away at the reproductive rights of women. And if those rights fall, other rights will follow. Their goal is to supplant modern society with a society that fits into their narrow world view. It all starts with an assault on *Roe*.

It's such a quiet assault that it is rare to even hear President Bush talk about women's reproductive rights. He doesn't talk about these issues to the media. He talks about them behind closed doors. He isn't proposing a constitutional amendment to ban abortion—a move that would tie him to the issue. He and Karl Rove have decided that they don't need to. Quietly. Without a lot of fanfare. They think they can accomplish their goals as the American public sleeps. While people toil away at their jobs and worry about their next paycheck, our opponents work. . . . This is the quiet front in the battle for our rights.

She then addressed several questions, all revealing her stridency on the issue. For example, she focused on abortion stances by "anti-choice forces" that "seem reasonable," but, in her view, are not. Among them, she noted, "It's a crime to harm a pregnant woman, so it should be a crime to harm the fetus, as well. Right? . . . We even believe in protecting the rights of doctors and nurses to act on their conscience in deciding what medical procedures to perform." She warned, "We should be careful in our complacency. Many of these policies sound perfectly reasonable to the untrained ear. But they are not reasonable when you realize the true intention—which is not to protect fetuses from crime, to expand access to prenatal care, to involve parents more thoroughly in their children's medical decisions, or to protect the civil rights of medical professionals. These policies are meant to chip away at *all* reproductive rights."

She then addressed another component in the abortion debate: the use of federal tax dollars to pay for abortions: "Anti-choice forces also argue that our tax dollars should not pay for abortion under any circumstances," said Mrs. Clinton. "On the surface, this argument also sounds reasonable. But by imposing this ban, Congress has denied access to a legal procedure for women who depend on the government for their healthcare—poor women, women in the military stationed overseas, Native American women, women in prison, federal employees, Peace Corps volunteers. These women are unable to make this deeply personal, most intimate decision even if they believe, and their doctors determine, it is in their best interest."

Senator Clinton not only made clear that Americans deeply opposed to abortion should hand over their tax dollars to pay for abortions, but portrayed these conscientious objectors as cruel, even bigoted toward poor and Native American women. Bringing class economics into the equation, she went on: "We have gone back in time to an era where the 'Haves' have a choice and the 'Have nots' are forced to rely on dangerous, if not illegal, procedures."

At this point in the speech, Mrs. Clinton let her emotions get the best of her and launched into an extreme assertion that pro-lifers were seeking to end "all rights of privacy." She stated:

> The truth is the pro-choice battle is just one example of their movement. Slowly, a powerful few are chipping away at much of the progress of modern society. Their first objective is to overturn *Roe*. To do that, they are willing to throw out all rights of privacy. Many of us say, "How can they so casually toss out the right of privacy to get at *Roe*? Don't they believe in privacy?" The answer is no, they do not. They simply do not believe in the right of privacy.

Though she was speaking to a receptive audience, these sweeping generalizations were extreme—even for Hillary. She did not provide examples, and in the months and years ahead she would find herself hard-pressed to account for the claim that pro-lifers seek to scrap "all" rights to privacy.

The speech continued to plummet downhill, descending into zealotry, as she began claiming that these same anti-choice forces—many of which, obviously, are scientists, doctors, medical professionals, people with doctorates and various other professional degrees—were opposed to both society and progress. Mrs. Clinton asserted:

In this society, progress has no place and science doesn't matter. In this society, fact is forgotten and evidence is ignored. All that matters is contained in their extremely limited world view. Any evidence that doesn't comport to their belief structure is tossed aside. Any law that doesn't agree with their belief is targeted for change. Any person who disagrees with them is labeled Un-American. . . . Evidence doesn't matter. Science doesn't matter. . . . Evidence doesn't matter, science doesn't matter, even the Constitution doesn't really matter. When the Constitution doesn't support their views, they say we should alter it, change it, reinterpret, until it fits in their world view.

That last line on reinterpreting the Constitution particularly angered pro-lifers, since, as everyone knows, including the most sympathetic pro-choice law professors, the right to an abortion had to be read into the Constitution, at the expense and total exclusion of sections guaranteeing a right to life, such as the Fourteenth Amendment, which states, "nor shall any State deprive any person of life." To the contrary, abortion was read into the "right to privacy," three words that themselves do not exist in the Constitution.[27] While Hillary must surely recognize these crucial constitutional facts, she overlooks them when seeking to rally her pro-choice base.

Mrs. Clinton continued this thought by angrily asserting, "What is so stunning to me is that these advocates of greater government power and reduced personal freedom can turn around and claim to be members of a political party that is supposed to favor a limited government—and they can do it with a straight face. We cannot and should not amend the Constitution to deny people freedom." Here, too, pro-lifers objected to her words; they believe that by denying women the "freedom" to have an abortion, they simultaneously preserve a more important, fundamental, and overriding freedom: the right to be born, the right to life. However, in this speech, Hillary did not entertain these distinctions. It was not a thoughtful speech; it assumed that "anti-choicers" were stupid, even vulgar, fascistic. The address demonized pro-lifers, and her audience loved it.

As she reached the crescendo of her address, Senator Clinton took her concerns global, warning her listeners: "This . . . has consequence for us, and for people around the globe. Think again about the beliefs of our opponents and its impact on reproductive rights." America's current leaders, said Mrs. Clinton, in a reference to the current Bush administration, were seeking to impose a "global gag rule" and "to close [reproductive] clinics and cut [reproductive] services" around the world. "These changes," she assured, "are wreaking havoc on women's lives." She gave examples from Kenya, Zambia, and Ethiopia. She again assailed the anti-choicers—for their intransigence, their fanaticism:

We can stand and tell our opponents of all the problems in the world because of uncontrolled population growth and unplanned pregnancy, but they won't listen. All the evidence in the world won't convince our opponents that they are wrong. All they need is their beliefs. With their beliefs in hand, they are chipping away at reproductive rights, at privacy rights, and at progress. That is their goal.

She then concluded with a clarion call, to rally the audience, to rouse her pro-choice allies into moving ahead with her in their shared crusade:

You—the people right here in this room—are the ones with the power to stop them. You have that power because you are the trained ear. You understand what their policies really mean. You understand the scope of this battle. And, you are not asleep. You have a great responsibility—to wake up this country.

It falls to you to make people aware that we are one Supreme Court nominee away from overturning *Roe v. Wade*; one Supreme Court Justice away from throwing privacy rights to the wind; one Supreme Court Justice away from a world where the beliefs of a few will dominate the many.

She then finished with a flurry, shouting at the NARAL audience: "Our rights are at stake. Our freedom is at stake. Our way of life is at stake. Let's wake up America!"

It was evident from this speech that no other issue so animated Mrs. Clinton. In none of her speeches on religion was she ever as zealous as in this one on abortion. Moreover, it would not be unfair to conclude from the text and tone of this speech that Mrs. Clinton seems to literally hate pro-lifers. In a separate context, she has confessed: "I wrestle nearly every day with the biblical admonition to forgive and love my enemies."[28] Yet with speeches such as this, it would seem to be a particularly acute challenge in the case of pro-lifers.

Finally, it must be noted that this speech on social policy by Mrs. Clinton contained no mention of God, Christ, or her faith, a common occurrence for Hillary, who has long been very deliberate in separating her faith from her position on abortion. As such, abortion is the one and only area where she insists that her faith does not influence her policy positions.

A few weeks later, Mrs. Clinton backed up her words with action. On March 25, 2004, she voted against the Unborn Victims of Violence Act. The bill would have allowed federal prosecutors to bring charges on behalf of a "child in utero" as a second victim when injured or killed during commission of a violent act against the mother. The bill was passed sixty-one to thirty-eight. All but two Republicans favored it, while most Democrats (thirty-five of forty-eight) opposed it, including Mrs. Clinton, as did one independent.

March for Women's Lives

In April 2004, Hillary took her NARAL speech a step further when she agreed to speak at the "March for Women's Lives" on the mall in Washington, D.C. Helping to kick off the event on April 23, which was highlighted by a Sunday, April 25, finale, was John F. Kerry, the Democratic Party's nominee for president in 2004. Kerry's opponent in November was not welcome at this event: Among the estimated half-million marchers—the National Organization for Women claimed the figure was 1.15 million[29]—were women carrying signs decrying "Bush's War on Women," and also condemning the president's mother, Barbara Bush, for not aborting her oldest son: "If Only Barbara Bush Had Choice," read one sign; "Barbara Chose Poorly," said another. Another prominent pro-lifer, Pope John Paul II, would not have received many votes at this rally either: "The Pope's Mother Had No Choice," lamented another sign.[30] A popular target at this rally was Christianity generally, and pro-life Christians in particular. Read one placard: "Pro-Life Is to Christianity as Al-Qaeda Is to Islam."

This is not to say that there were no religious speakers or religious imagery at the event. Planned Parenthood sent its official chaplain. Also in attendance were the Christian Dykes for Choice, the Religious Coalition for Reproduction Choice, and Catholics for a Free Choice, the last of which was represented by Francis Kissling, who described the gathering on the mall as a "sacred place . . . the place to be, not the churches." Likewise, the Reverend Barry Lynn of Americans United for Separation of Church and State described it as "hallowed space." Amid placards with statements like, "This Is What a Jewish Feminist

Looks Like" and "Episcopalians for Choice" and "I Asked God, She's Pro-Choice," were chants such as "Tax the Church! Tax the Church!"

There was actually a notable degree of religious talk at the event. One of the featured speakers, Representative Maxine Waters (D-Calif.), shouted from the podium that George W. Bush needed to "go to hell." This was not an isolated thought that weekend, as one of those who followed Waters fumed against the "Bush/Satan administration," which must have been a corrective to the congresswoman, implying that either Bush had already been to hell or that Satan had come up—either way, an alliance reportedly had been agreed upon. Another speaker felt that the "religious right" was an inaccurate term for a movement that instead, she averred, ought to be called the "religious reich." A female rabbi said that to be "pro-choice" was to be "pro-God." The infamous abortion doctor George Tiller—whom pro-lifers call "Tiller the Killer"—was even in a preaching mood; invoking the Book of Revelation, he referred to the unholy alliance of George Bush, Dick Cheney, Donald Rumsfeld, and John Ashcroft as "the four horsemen of the apocalypse."

It was into this zoo that Mrs. Clinton stepped (as did former Clinton secretary of state Madeleine Albright). It was a Sunday morning, but the crowd was unlike anything attending a Billy Graham crusade. As Mrs. Clinton took the stage, said a reporter from *The Nation*, "it was as if the Beatles arrived." Young ladies cheered and screamed.[31]

Hillary told the crowd that the last time abortion rights advocates rallied in Washington, in 1992, the country had elected her husband as president, and abortion had been saved. Thus, she said, for twelve years there had been no reason to march, because there was a president and an administration, she said, "that respected the rights of women."[32]

Those abortion rights were now under assault by the Bush administration, a fact that enraged the speaker and the crowd. Here again, as in her NARAL speech four months earlier, Hillary's tone was one of vituperation. She said the Bush administration was filled with people who "disparage sexual harassment laws," who believe that the "pay gap" between men and women is a farce, and who believe (she was correct here) that *Roe v. Wade* represents "the worst abomination of constitutional law."[33]

Clinton urged the ladies to vote the Democratic Party line in November. The entire event, reported one source, was "a pep rally for Democrats."[34] Mrs. Clinton said that the March for Women's Lives would be in vain if the women did not show up at the polls in the fall. She urged the crowd to elect a pro-choice president.[35] When she finished, the emcee prophesied, "Thank you, *President* Clinton." The crowd exploded.[36]

Again in a place where religion seemed to hold something of a dubious reputation, Hillary steered clear of the subject altogether. Despite her respect for the Christian faith, she remained silent amid the sea of rather uncharitable signs and slogans that seemed to misuse or even abuse religious imagery and individuals.

Values Voters in 2004

With its heavy rhetoric and rallying images, Hillary's speech at the March for Women's Lives created a marked contrast with her tone in the aftermath of the November elections. On November 2, 2004, George W. Bush handily won reelection over the Democratic candidate, John Kerry, winning more than 50 percent of the popular vote, a milestone Bill Clinton never reached. He did it with a big boost from churchgoing Christians, the so-called values voters who were crucial to his victory. Indeed, the 2004 vote was a wake-up call and potential watershed for a religious Democrat like Mrs. Clinton.

A telling difference between Kerry and Bush, which was crucial to these values voters, was how their faith related to their positions on abortion. Bush said that a life in the womb is a gift from God that should be protected, and his actions in office backed up this position as he sought to overturn Bill Clinton's policies from the 1990s. Kerry's position was more complicated. In the final presidential debate on October 13, 2004, he said, "My faith affects everything I do, in truth. . . . And I think that everything you do in public life has to be guided by your faith, affected by your faith, but without transferring it in any official way to other people." He explained that this credo explains "why I fight against poverty," "why I fight to clean up the environment," and "why I fight for equality and justice," all of which he as a legislator transfers in an official way to other people. However, like Hillary, the only area where Kerry seemed not to allow his faith to influence his public life was abortion.

As president, John F. Kerry would have changed the direction that Bush had been heading in, pledging to fill any openings in the U.S. Supreme Court with pro-choice appointments. Like Mrs. Clinton, he was one of the most inflexibly pro-choice members of the Senate, and a popular speaker at events like NARAL gatherings and the March for Women's Lives. At the 2003 NARAL Pro-Choice America Dinner, where, in language much like Senator Clinton's, he described pro-lifers as "the forces of intolerance," Kerry boasted that his maiden speech as a freshman senator had been in support of *Roe v. Wade*. On the Senate floor, he stated: "The right thing to do is to treat abortions as exactly what they are—a medical procedure that any doctor is free to provide and any pregnant woman free to obtain. Consequently, abortions should not have to be performed in tightly guarded clinics on the edge of town; they should be performed and obtained in the same locations as any other medical procedure. . . . [A]bortions need to be moved out of the fringes of medicine and into the mainstream of medical practice."[37]

In April 2004, Kerry took a rare timeout from the presidential campaign to appear on the Senate floor to join Mrs. Clinton in voting against the Unborn Victims of Violence Act that would make it a crime to harm a fetus during an assault on the mother. He also joined Senator Clinton in voting against the ban on partial-birth abortion.

These pronounced differences between John Kerry and George W. Bush on abortion mattered significantly in November 2004, as Mrs. Clinton soon learned: According to CNN exit polling data, 22 percent of the electorate, or 25.3 million voters, said that "moral values" was their most important issue—79 percent of them cast ballots for Bush.[38] Over a quarter of all voters, 26 percent, said that abortion ought to be "mostly illegal," and they went for Bush by 72 percent to 27 percent, or by 21.5 million votes to 8.1 million votes. Those 15 percent who said abortion should be "always illegal" cast a ballot for Bush by 77 percent to 22 percent, or 13.3 million to 3.8 million, a Bush advantage of nearly 10 million votes. This surge of pro-lifers was motivated and mobilized by Kerry's stridency on the abortion issue in a way unseen since the *Roe v. Wade* decision was handed down.

Equally interesting, and related, were the election figures based on church attendance: According to the same CNN exit poll data, those who attend church more than once per week made up 16 percent of 2004 voters, or 18.4 million voters, and they went for Bush by 63 percent to 35 percent, or by 11.6 million to 6.4 million, a difference of 5.2 million votes. On the other hand, being perceived as a secular or even unreligious candidate had some political traction in November 2004: Those who said they never attend church, which equaled 15 percent of voters, or 17.3 million voters, went for Kerry by 64 percent to 34 percent, or by 11.1 million to 5.9 million, also a difference of 5.2 million votes. These numbers were similar to the 2000 vote, when those who attended church more than weekly went for Bush by 63 percent to 36 percent, whereas Vice President Al Gore won those who never attended by 61 percent to 32 percent.

Ten percent of those who voted on November 2 claimed no religion at all. They made up nearly 15 million voters. Of those, 68 percent, or 10.2 million, voted for Kerry, but only 30 percent, or 4.5 million, voted for Bush—a Kerry advantage of 5.7 million votes. In other words, those religious voters often credited for winning the day for

George W. Bush in the 2004 presidential contest were countered by nonreligious Americans who tried to win the day for John F. Kerry.

The non-churchgoing vote was even larger in states where Kerry got the most ballots. In California, 24 percent of voters, almost one in four, said they never attend church, and they went for Kerry 63 to 34 percent. In New York, those who claimed no religion at all voted for Kerry by 78 percent to 19 percent. These eager atheists comprised 12 percent of New York voters, and they offset those Catholics in New York who favored Bush by 51 percent to 48 percent.

The Catholic Vote

Yet, even as many of these numbers offset each other, they signaled an important fact to Democrats in 2008, one noticed right away by Mrs. Clinton: If a Democrat can peel off a sizable sliver of these religious voters, while continuing to win the nonreligious vote, that Democrat can sink the Republican nominee. This is especially true when analyzing the Catholic vote in 2004.

It was frequently asserted that George W. Bush won the 2004 presidential election because of religious voters, especially evangelical Protestants. What was not commonly noted was that John F. Kerry lost the election because he failed not only to win religious voters generally but Catholics specifically. Kerry lost because he lost Catholics, an amazing fact when one considers that Kerry himself is Catholic. Kerry lost the Catholic vote to Bush by at least a million. A Catholic with a major party nomination traditionally would have won the Catholic vote by several million. Decades earlier, another Democratic senator from Massachusetts with the initials "JFK," John F. Kennedy, won an extremely close election because he overwhelmingly took the Catholic vote.

Put differently, Catholics voted for the Protestant, George W.

Bush, and did so in large part because they agreed more with him than Kerry on moral issues, rooted in religion, that were closest to Catholic hearts, like abortion. Just as Al Gore did not win the electoral college in 2000 because he could not carry his home state of Tennessee, John Kerry failed because he could not bring along a natural constituency. According to CNN's exit poll data, 27 percent of those who voted that first Tuesday in November were Catholic, which equated to roughly 31 million of 115 million voters. How these Catholics voted is striking: They voted for Bush over Kerry by 51 percent to 48 percent. In other words, they mirrored the popular vote to the exact percentage.

The numbers diverge more sharply when one compares devout Catholics to those who find their way to church only for weddings and Christmas: Catholics who attend Mass weekly voted for Bush by 55 percent to 44 percent, a startling religious rejection of John Kerry. The breakdown among states is most interesting: Bush remained close to Kerry in Pennsylvania—a state that has millions of pro-life Catholic Democrats, and that went for Kerry 52 percent to 48 percent—in large part because Bush carried Pennsylvania Catholics who go to Mass weekly by 52 percent to 48 percent. In New Hampshire, which barely went for Kerry, Bush took Catholics who attend Mass weekly by 63 percent to 35 percent.

Most impressive, Catholics played a key role in Florida and Ohio, the two states watched most closely in 2000 and 2004, respectively, and the two that were called late and made the decisive electoral college difference for Bush. In Florida, Catholics comprised 28 percent of voters, and went for Bush 57 percent to 42 percent. In Ohio, they made up 26 percent and went for Bush 55 percent to 44 percent. The margin was even wider for Catholics who attend Mass weekly: In Florida, they went for Bush by almost two to one, 66 percent to 34 percent, and in Ohio they supported Bush by 65 percent to 35 percent.

In fact, Catholics for Bush made it unnecessary to begin counting provisional ballots in Ohio. Ohio Catholics cast 780,000 votes for Bush and 624,000 for Kerry, a difference of 156,000 votes. Compare that figure to the overall vote difference for all Ohio ballots, which was 136,000. Thus, it can be asserted that John Kerry lost Ohio, and thus the election, because he could not get the support of people of his own faith in Ohio.

The issue behind this Catholic snub was abortion. Pro-life Catholics were aghast at the prospect of a Catholic president becoming the greatest champion of legalized abortion ever to step foot in the Oval Office, as Kerry would have been. Kerry protested by pointing to social justice: His piety would prompt him to boost the minimum wage and clean up the environment, but ultimately these were not the social justice issues that Catholics were voting on.

While in the aftermath of the election the common explanatory refrain was that Bush political strategist Karl Rove secured the religious vote by mobilizing evangelicals, the fact is that the Democratic Party, by running John F. Kerry, drove both evangelicals and Catholics toward Bush. Kerry did more for Protestant-Catholic unity in America than the churches themselves could accomplish. In truth, moderate to conservative Catholics had nowhere to go but to George W. Bush, even as many had grave reservations over the war in Iraq.

CNN's 2004 exit poll data point to this statistically accurate profile: Next to African Americans, the surest Democratic voter was an unmarried, city-dwelling, Northeast, pro-choice atheist with a graduate degree who thinks that gay people should be legally married. The least likely Democratic voter was a married citizen who regularly attends church, thinks abortion should be illegal, and cites "moral values" as a top priority.

The Democrats React

To be sure, the Democrats were not taken totally by surprise. The 2000 data had been nearly identical. The Democratic Party knew that it would need to make inroads with religious and values voters in 2004. Thus, John Kerry talked several times about his faith, and even got quite aggressive: Picking up the mantle from Al Gore, Kerry did what Democrats are able to get away with but Republicans cannot: He used his faith to question Bush's. Throughout the campaign, Kerry employed a Bible verse to question the Christian commitment of his political opponent, using the New Testament's James 2:14, which asks: "What good is it, my brothers, if a man claims to have faith but has no deeds?" Kerry applied this verse to doubt whether the president lived out his faith and whether Bush's policy choices were rooted in Christian principles.

Kerry began doing this early in the campaign, much earlier than had a desperate Al Gore in the waning weeks of 2000. Kerry started the tactic in a March 7, 2004, speech at a Mississippi church and leveled the accusation again on March 28, speaking at a Baptist church in St. Louis. After these first attempts resulted in no negative attention from the press, Kerry did not relent, aiming the passage at Bush too many times to count. In the final presidential debate, nationally televised on October 13, he twice applied the verse to Bush.

James 2:14 was John Kerry's most frequently quoted Bible verse on the campaign trail, cited not to illuminate his own faith but to directly question the sincerity of Bush's. Despite all the wild accusations about George W. Bush's tactics, he never stooped to the level where he publicly questioned Senator Kerry's faith. As he knew, if he did, he would have been pilloried.

Like Al Gore in 2000, Kerry got assistance from some special friends on this front in 2004: Bill Clinton, who witnessed the train wreck in November 2000, sought to help head off disaster in 2004.

In the most underreported story of the political season, as the New York papers tripped over themselves to find out if Bush's chief political operative, Karl Rove, was trying to hunt down registries of evangelical churches for the purpose of driving votes, Bill Clinton at the start of the Republican convention gave a homily at the radical Riverside Church in New York. Clinton addressed the congregation during the worship service. He accused Republicans of bearing "false witness" and being "the people of the Nine Commandments." The pastor introduced Clinton as part of an announcement of the church's Mobilization 2004 campaign. The *New York Times* and *Daily News* were not interested in this story: They devoted total attention to the Rove story.[39]

Al Gore himself lent a hand: In an October 18 speech at Georgetown, the former vice president ripped into the current president: "I'm convinced that most of the president's frequent departures from fact-based analysis have much more to do with right-wing political and economic ideology than with the Bible."

Despite these claims of being "holier than thou," the election still resulted in defeat for the left. In the days and weeks after, as strategists and analysts sifted through the numbers and drew conclusions about the future of the Democratic Party, the reality for Democrats was unassailable and unavoidable: Regardless of Kerry's religious values, it was not enough to be comfortable talking about God and religion. Furthermore, as the failure of Kerry's James 2:14 quote demonstrated, it was not enough to denigrate the faith of the other candidate. Presidential candidates who want to win a general election need to back up their religious positions with stances, actions, and programs for the moral, values-based issues that drive faith-inspired voters to the polls.

These conclusions posed intriguing questions relating to Hillary: Neither Gore, Bill Clinton, nor Kerry had any credibility with the values voters they were trying to move—but did Hillary have that authority? And, if not now, could she someday? Like her husband after the 1994 vote ten years earlier, Hillary understood the larger implications of the situation immediately and was among the first to react to this religious verdict. She immediately understood the faith factor in the 2004 vote.

This was evident mere days after the election, during a speech she gave on November 10 at Tufts University. In her address, she called it a mistake for Democrats to have not engaged evangelical Christians on their own turf, thereby ceding the vote to President Bush. "I don't think you can win an election or even run a successful campaign if you don't acknowledge what is important to people," she said, referring to the importance of faith. "We don't have to agree with them. But being ignored is a sign of such disrespect. And therefore I think we should talk about these issues."

She then singled out areas where she thought faith-backed Republican politicians were vulnerable, pointing to social justice: Hillary said the Bible should be cited to win debates over poverty, akin to how Republicans referenced Scripture to resist the legalization of gay marriage. "No one can read the New Testament of our Bible without recognizing that Jesus had a lot more to say about how we treat the poor than most of the issues that were talked about in this election," said Senator Clinton, suddenly sounding once again like the Methodist from Park Ridge.[40]

To Hillary, the objective lesson here was not how Jesus felt about abortion, but how Jesus felt about the minimum wage. That, she judged, was a winning strategy. But irrespective of the specific issue at stake, the tone of this speech differed greatly from that of her speeches during the buildup to Election Day. The speech itself was quite removed from her battle cry at NARAL and the March for Women's Lives. Her fiery language had been replaced by words of consolation and outreach. Suddenly her speech was overflowing with talk of the middle, of values that every Christian American deemed important. Of course this is not to say that her positions seemed to have changed at all, simply that her phrasing of them had altered.

Only weeks later, Mrs. Clinton got great news. In mid-December came the results of a stunning poll by Fox News/Opinion Dynamics: If the 2008 presidential election were held that day, Senator Clinton would handily defeat three of the top Republicans being touted as possible candidates, beating Senate Majority Leader Bill Frist (R-Tenn.) 40 percent to 33 percent, taking New York Governor George Pataki 41 percent to 35 percent, and easily dispatching Florida Governor Jeb Bush 46 percent to 35 percent. The Fox News survey found that a "clear majority" of voters (59 percent) judged Hillary qualified to be president.[41]

The poll offered the momentum she needed. Now, looking ahead to 2005, Senator Clinton began seeking ways to appeal to the political middle, especially those religious voters. That included fixing her gaze upon that crucial Catholic constituency.

CHAPTER 14

Moving to the Middle

By January 2005, Mrs. Clinton was into the fifth year of her Senate term. That month presented a fascinating contrast in her new strategy to attract a broad range of voters in the run-up to her candidacy for the 2008 presidential election, but it also highlighted the risks that she runs as she seeks to maneuver through some of the country's most divisive religious issues.

On January 11, 2005, Senator Clinton gave the keynote to the International Women's Health Coalition Fourth Annual Gala. Titled "Meeting Global Challenges: Healthy Women, Healthy World," it was a speech that proved to be as open and forthcoming about her pro-choice stance as any that she had given. Citing her work with Senator Barbara Boxer, Mrs. Clinton spoke of her proposed amendment to the Global AIDS Bill, which sought to provide assistance to foreign countries to combat HIV, tuberculosis, and malaria, "and in part to provide comprehensive assistance for programs for women and girls."

The latter meant abstinence programs, condoms, and abortion clinics. "Today, an estimated 20 million women worldwide risk unsafe abortions every year," she said. "And yet, as we know, the current administration is making it more difficult for women to receive the full range of health services." She noted that "under the global gag rule," reinstituted by the Bush administration, no U.S. funds could be provided to foreign nongovernmental organizations that provide abortions or advocate abortion counseling or legalization. Seeming to pick up lines from many of her preelection criticisms of Bush, she condemned "these ill thought out policies by this administration." She also issued a significant public reaffirmation that "reproductive health care and family planning service is a basic right"—abortion is a basic right.[1] Catholics, by contrast, have called the right to life "the most basic right."

No doubt, those fresh words from Senator Clinton were still ringing in the ears of those who had just learned that Canisius College had invited her to speak. On January 18, 2005, Canisius, a small Catholic college in Buffalo, New York, issued a press release announcing that it had invited Senator Clinton to speak in its lecture series on the "Governmental Role in Effectuating the Corporal Works of Mercy," to be delivered on Monday, January 31, at 1:15 P.M. in the college's Montante Cultural Center, cosponsored by the college's Committee for the Promotion of Justice. The former first lady would address government's role in caring for the sick. As a member of the Senate committees for environment and public works, and health, education, labor and pensions, noted the college's press release, Senator Clinton had introduced legislation to expand health care coverage to children and aging Americans. The press release added that Canisius College, one of America's twenty-eight Jesuit colleges and "the premier private college in Western New York," specialized in preparing leaders—"intelligent, caring, faithful individuals"—who would "promote excellence in their professions, their communities and their service to humanity."[2]

The little Catholic college was proud of its celebrity event, but many Catholics who were not responded by rapidly organizing boycotts. The Buffalo chapters of Catholic Charities and the Office of Church Ministry withdrew their sponsorship, and the Buffalo diocese received hundreds of calls and e-mails complaining about Clinton's appearance.[3] The Buffalo Regional Right to Life Committee was outraged, accusing Canisius officials of creating "a grave scandal within the Catholic community" by inviting Mrs. Clinton.[4] The national Catholic press joined the chorus, including the popular Web site Catholic Exchange, which dubbed Hillary the "abortion crusader."[5]

Critics pointed to guidelines by the ten American Jesuit provincials and a letter by the U.S. Conference of Catholic Bishops stating that those who violate Catholic principles "should not be given awards, honors or platforms." The bishop of the Buffalo diocese, the Most Reverend Edward U. Kmiec, said that he was not pleased that Canisius College invited Clinton, especially without consulting him in advance, but he declined to order Canisius to ban the pro-choice former first lady from campus.[6]

The bishop did, however, withdraw the Diocese of Buffalo's sponsorship of the speech, and issued a stern statement: "As Bishop of the Diocese of Buffalo, I wish to inform the faithful and the community that the Diocese of Buffalo is not associated with the planning or promotion of the lecture of Senator Hillary Clinton at Canisius College on Monday, January 31, 2005. This event has been arranged under the auspices of Canisius College without previous consultation." He also reported that Catholic Charities had been listed as a sponsor without prior knowledge that Senator Clinton would be one of the presenters of the lecture series, and had now also withdrawn sponsorship. "We have communicated our displeasure," said the bishop.[7]

Ten-Point Leadership Foundation

The day after the Canisius announcement, Senator Clinton regrouped. On January 19, 2005, the eve of George W. Bush's second inaugural,

Mrs. Clinton embraced an issue that was important to the religious voters who in November had helped elect George W. Bush: faith-based solutions to social problems. She addressed a fund-raising dinner in Boston for the National Ten-Point Leadership Foundation and the Ella J. Baker House, two groups that are big boosters of faith-based initiatives—the cornerstone of George W. Bush's compassionate conservatism and integration of faith and politics.

"There is no contradiction between support for faith-based initiatives and upholding our constitutional principles," declared the senator in an important line bound to be quoted in the years ahead. She said there was a "false division" between faith-based solutions to social problems—backed by the public sector—and the separation of church and state. In a talk that invoked God at least a half-dozen times, she emphasized to the audience of more than five hundred, including many religious leaders, that "I've always been a praying person." The senator from New York insisted to the Boston crowd that faith-filled individuals be permitted to "live out their faith in the public square."[8]

George W. Bush could not have agreed more. Had he been there, he would have been on his feet. It was his type of program—and his type of group, directed by the Reverend Eugene F. Rivers III, a leader among clergy to halt crime and violence among city youth, and who has been critical of Boston's liberal politicians for separating faith from the policies that work effectively. Clinton praised Rivers and his (as Bush calls them) soldiers in the army of compassion. Rivers and his team, said Hillary, could "see God's work right in front of them."

Hillary's display of her religious credentials left many suspicious of her timing. Among many Democrats, Bush's support for faith-based organizations had been one of his most suspect domestic initiatives. Now suddenly in the wake of Republican victories, Hillary was trumpeting their value and insisting on their importance. These programs were precisely the types of ideas that made Bush popular with religious Americans as he sought to bring faith into public life. Indeed, support for faith-based organizations is a home run with Catholics and other Christians, including the likes of James Towey, a former twelve-year legal counsel to Mother Teresa, who lived as a full-time volunteer in her Washington, D.C., home for AIDS patients, and who became the director of President Bush's pet project, the Office of Faith-Based and Community Initiatives. These were precisely the people that Hillary would need to win over in her eventual run for president. At the same time, the public should not have been too suspicious: Hillary had spoken to audiences about her faith since the Arkansas days. This was not entirely new political terrain for her.

The Albany Speech

The Canisius fallout, on the heels of the November vote, had taught Hillary something significant about her vocal pro-choice stridency, as evidenced not even a full week later, when she was scheduled to provide remarks to another fiercely pro-choice crowd, the annual conference of the Family Planning Advocates of New York State.

Gathering in Albany, the faithful expected another no-holds-barred talk framing "anti-choicers" as intransigent fanatics conspiring to rip away all privacy rights. What they got was something quite different, as a much more politically savvy Hillary did not explode into rage against "anti-choice forces" as she had twelve months earlier. The zealots of last January were the Canisius protesters of this January and the values voters who had defeated John F. Kerry in November. Meanwhile, the press did its part to cite the speech over and over as evidence—which it was not—that Mrs. Clinton was "moderating" her stance on abortion. In fact, as the below excerpts make clear, she was moderating her rhetoric toward pro-lifers, while not changing her fervency for legalized abortion in a single area of policy.

She actually began the speech much like the NARAL talk: "I am so pleased to be here two days after the thirty-second anniversary of *Roe v. Wade*, a landmark decision that struck a blow for freedom and equality for women. Today *Roe* is in more jeopardy than ever, and I look forward to working with all of you as we fight to defend it in the coming years." Just then, however, the tone changed right away, with an olive branch rather than a grenade to her opponents: "I'm also pleased to be talking to people who are on the front lines of increasing women's access to quality health care and reducing unwanted pregnancy—an issue we should be able to find common ground on with people on the other side of this debate." The words "common ground" rippled well beyond that auditorium; they were two of the most enduring words to emerge from her speech.

She did once again castigate the current administration for its international policies: "I heard President Bush talking about freedom and yet his administration has acted to deny freedom to women around the world through a global gag policy, which has left many without access to basic reproductive health services." But in the next line came another statement no one expected: "I believe we can all recognize that abortion in many ways represents a sad, even tragic choice to many, many women."

That would be the most quoted line from the speech. Truthfully, it was a statement of the obvious: Hadn't Al Gore and even John Kerry said that abortion was regrettable? Yet it was a big deal to hear it from Mrs. Clinton, and to hear it in this setting where people saw abortion as an unassailable absolute, never to be questioned in any way, or even to be acknowledged for the physical and emotional pain and problems it so often caused.

She then reached for more middle ground, noting, "Research shows that the primary reason that teenage girls abstain is because of their religious and moral values. We should embrace this." Continuing, she extended another olive branch to pro-lifers, who, suddenly, were no longer the narrow-minded, simplistic "anti-choicers" they had been a year earlier—quite the contrary. Mrs. Clinton, in a startling about-face, now valued their opinions: "I for one respect those who believe with all their hearts and conscience that there are no circumstances under which any abortion should ever be available."

This speech marked a rare occasion whereby Hillary had allowed the idea of religious and moral values to infiltrate one of her speeches on abortion. It was notable for many reasons, but mostly because it presented such a striking dichotomy from the previous year. The contrast in style, the lack of combativeness and contempt, compared to the NARAL address, was absolutely remarkable, an unmistakable example of Mrs. Clinton modifying her approach for political purposes. The 2004 election was the only intervening event that could explain the turnabout in her posturing; Hillary did not want to see a repeat of 2004 in 2008, if and when she became the Democratic nominee. Demonizing pro-lifers worked beautifully in New York City, but would fail miserably in Ohio.

The *New York Sun* rightly editorialized the next day, "No one who listened to Senator Clinton's speech to abortion rights supporters at Albany yesterday can have any doubt that she's running for president."[9] This was obvious to everyone—well, almost everyone. The *New York Times* responded by running with the issue in precisely the way the senator's staff had hoped. "These are practical steps for cutting the nation's abortion rate," wrote the *Times* in an editorial that commended "Mrs. Clinton's frank talk."[10]

Philippe Reines, Hillary's press secretary, must have read the *Times* editorial three or four times, each time with a bemused chuckle. "The times may have changed, but her beliefs have not," he said, stating the obvious. The times dictated that she change her rhetoric, but her heart insisted she not modify her position.[11]

For its part, the pro-choice lobby was not going for this olive-branch nonsense, and it was not going to accept Hillary's characterization that there was anything sad or tragic about abortion. One of the attendees, Martha Stahl, director of public relations and marketing for Northern Adirondack Planned Parenthood, told the *New York*

Times: "[W]e see women express relief more than anything else that they have the freedom to choose." Likewise, Ron Fitzsimmons, president of the National Coalition of Abortion Providers, stood firm and proud: "We have nothing to hide. The work we're doing is good." Abortion, they insisted, should not be stigmatized.[12]

Arrival at Canisius

The senator hoped that the change in tone would make for an easier landing in Buffalo, where, on Monday, January 31, 2005, the speech at Canisius went forward. Yet pro-life Catholics were not mollified by the moderation in tone. Mrs. Clinton's presence drew protesters from several Catholic, pro-life, and Republican groups, including Judie Brown, president of the American Life League.[13] "Sen. Clinton's entire career has been characterized by complete defiance toward the Church's stance on the sanctity of innocent human life," Brown told the press. "Her repeated statements that *Roe v. Wade* should remain the law of the land illustrate how fundamentally opposed she is to the Catholic Church on this most basic of civil rights."[14] Brown added that "hosting a pro-abortion speaker such as Sen. Clinton is simply not in keeping with the teachings of the Catholic Church."[15]

Apparently, common ground had not been found. This was most certainly not what Mrs. Clinton had envisioned when she planned to reach out to disgruntled Catholics who rejected John Kerry two months earlier, and as she was quickly learning, appealing to Catholics would be much harder than she had imagined. Social justice alone was not enough, regardless of what advisers and friends on the religious left had told her.

While Hillary went on to make the speech, the extent of the furor that it created within the Catholic community prevented her from gaining any true traction on the issue. Still, she was trying to make inroads into unfamiliar territory, and these inroads would take work.

Some of these experiments would prove too unfriendly and too controversial to make a difference. Others would be met by friendly press and friendly moderates who felt that this change in language signaled a new side to the Democratic senator. Regardless of the outcome of these specific events, Hillary found herself with the luxury of time. Though she was widely suspected to be the Democratic Party's leading contender for 2008, the primaries themselves were still three years away. Until then, she had time to refine her strategy and cultivate relationships.

Right to Live and Die

Symbolic of Hillary's struggles to repackage herself as middle ground on issues of religion and morality was her reaction to the demise of the world's preeminent pro-life Catholic, Pope John Paul II, with whom the Clintons had butted heads during the presidential years, and another Catholic who was dominating headlines in the United States: Terri Schiavo.

A battle was raging between what the pope called the Culture of Life and the Culture of Death over whether to remove a feeding tube from Schiavo—her husband's wishes. Her parents insisted she was not brain-dead and that they be permitted to take care of her. Hillary, whether thinking of politics or maybe the death of her father a decade earlier, stayed out of the issue, leaving it to the Schiavo family, even as it came before Congress.[16] A Florida judge, granting the request of Schiavo's husband, ordered that her feeding tube be removed, and that doctors permit her to be starved to death.

The underwhelming nature of Hilary's response to Schiavo left some of her liberal supporters fit to be tied. Syndicated columnist Molly Ivins wrote an op-ed titled, "I Will Not Support Hillary Clinton for President."[17] The reason, wrote Ivins, was Mrs. Clinton's "failure to speak out on Terri Schiavo." Indeed, as the controversy heated up,

Hillary failed to take an open stance on the issue that rapidly became one of the country's most talked about debates. As religious, cultural, medical, and political groups from around the country weighed in with their support for either Schiavo's husband or her parents, Hillary kept to herself, evidently unwilling to tread into the political hot water that the discussion entailed.

In the end her silence proved shrewd; had she spoken in support of Schiavo's parents, activist groups on the left would have no doubt vilified her as a traitor. On the other hand, if she had sided with Schiavo's husband, she would have been opening herself up to criticism from the right-to-life movement that she had been arduously courting since Election Day the previous year. It was a precarious tightrope, and the only way to walk it was to remain silent.

Meanwhile, as Schiavo began to enter the point of no return around March 30, a dying, suffering John Paul II himself had a feeding tube inserted and clung to life, barely reaching the period that formally begins Divine Mercy Sunday, a feast day created when he canonized a little-known, uneducated Polish nun in 2000. He died on April 2, barely outliving Schiavo, and no doubt hoping to provide America and the world with a stark contrast and a reminder of the value of life.

On Sunday, April 3, just hours after the death of the pope, both Hillary and Bill jointly praised the man as a "beacon of light" for all people and a "force for democracy." Bill said that he and his wife were "deeply saddened" by the passing. In a separate statement from her Senate office, Hillary praised the pontiff's "authority as a force for democracy, tolerance, and forgiveness," commending his "unmatched intellect, his infinite heart," and his "message of love and hope."[18]

Mrs. Clinton was very careful with her words; her husband, however, was not. As the Clintons were aboard Air Force One, at the invitation of the current president, to fly to Rome to attend the pope's funeral, Bill was asked about the pontiff's legacy. Eager to talk about himself, he drew comparisons between the pope's legacy and his own, telling reporters that John Paul II "may have a mixed legacy," a comment often made about Clinton. This prompted the current president, George W. Bush, to counter that the pontiff was leaving a "clear legacy," a "strong legacy of setting a clear moral tone."

Bill's comments were a bit clumsy, and highly offensive to some, even though, when viewed in full, they were not that bad: "There will be debates about him," explained Clinton. "But on balance, he was a man of God, he was a consistent person, he did what he thought was right. That's about all you can ask of anybody."[19] Basically, Clinton had spoken for millions of liberals who felt the same way.

Nonetheless, outraged conservatives launched straight to the moon: The pro-life Web site LifeSiteNews.com may have gone over the top when it concluded that Clinton, who, it noted, was "one of recent history's most ardent political advocates of abortion," had "placed himself on equal footing with the Pope." The words of Pastor Joseph Grant Swank Jr. of the New Hope evangelical church in Maine were blistering: "This rank sinner of the most alley cat genre plows into the Pope on the way to the Pope's funeral!"[20] According to another detractor, columnist Michael J. Gaynor, Clinton was lucky that "a merciful God did not strike him dead on the spot" for the comparison.[21]

Hillary, as usual, had the better sense to avoid any self-comparisons, as did George W. Bush, who kept a low profile in Rome. "He recognized the significance of the moment," said Bush spokesman Scott McClellan. "And the focus should rightly be on the Holy Father." As McClellan made that statement, Bill Clinton had just finished an interview on the pope with *NBC Nightly News* anchor Brian Williams, watched by millions.[22]

Much can be said of the two deaths in Rome and Florida that week, particularly in regard to the Clintons' position on "life" issues like abortion and euthanasia. The deaths of Pope John Paul II and Terri Schiavo at that particular time, during that week, and the opposing ways in which they occurred, were viewed by many Christians as not coincidental. Recall the pope's words directed at the crowd and the Clintons in Denver on August 12, 1993: He had said that the right to life and protection of the human person needed to be the "great cause" of the post–Cold War United States. It was this bridge between the end of Communism at the close of the twentieth century and the cause of life so prominent at the start of the twenty-first century that the pope's own death—in keeping with his words—now seemed to represent. With the death of Pope John Paul II, it seemed that the twentieth century had finally ended, a tad late, and the twenty-first century had, alas, only now begun.

Despite the new image that she was trying to cultivate, it seemed unlikely that Hillary agreed—if she did, she certainly did not say so. Fittingly, it was these very same life issues that continued to hound Mrs. Clinton in her bid to position herself as a moderate and a Christian Democrat.

On April 25, the pro-life Christian Defense Coalition fired off a smoking press release: In response to Hillary's olive branch in her speeches, the Christian Defense Coalition—hoping to spark a dialogue and find common ground with Hillary—maintained that it had tried throughout February, March, and early April to arrange a face-to-face meeting with Senator Clinton, who ultimately declined.

"We are profoundly disappointed that Senator Clinton has refused to sit down and dialogue with a broad coalition of voices within the pro-life community after suggesting she wanted to reach out to us," said the Reverend Patrick J. Mahoney, director of the coalition. "It now seems that the statements Senator Clinton made, concerning finding common ground on abortion, were politically motivated and not sincere. These comments should now be viewed as an attempt to reinvent herself and appear less radical on the issue of abortion in light of the 2008 race for the White House." Mahoney cautioned: "I think it is critical that the American public focus more on what Senator Clinton does concerning abortion, rather than her rhetoric, as we move closer to the presidential campaign season."[23] The coalition held a news conference on Tuesday, April 26, outside the Russell Office Building on Capitol Hill.

If she was actually looking to find "common ground" with elements within the pro-life movement, ignoring the Christian Defense Coalition was a wasted opportunity. While it is far-fetched to think that the two sides would have shared any specific policy goals about abortion procedures themselves, the mere act of sitting down at the table together might have had a lasting effect, as it would have backed up her words with actions. The two sides could have discussed a more benign subject, such as teen abstinence policy, as a way to begin a dialogue, much in the way that Hillary had with Mother Teresa a decade earlier. An act like that would have gone a long way to proving that there was significance simply in discussion and her willingness to put her religious values at the forefront of her policy. Unfortunately, the group's entreaties were met with silence.

Public Religion

While the Christian Defense Coalition was questioning Hillary's sincerity on certain values, Hillary tapped other ways to showcase her faith for religious voters. In April 2005, Hillary took her faith credentials in a different direction by cosponsoring the Workplace Religious Freedom Act. For this piece of legislation, she joined forces with conservative colleague Rick Santorum (R-Pa.) to write a bill that guaranteed the right to religious expression on the job without fear of recrimination. This meant, for example, that an Orthodox Jew who honors the Sabbath cannot be forced to work on the Sabbath against his or her will, or that a Christian can wear a crucifix, or that a Sikh can don a turban. Backers of the bill included a broad coalition of forty clerics representing nearly every major denomination.

Hillary's support of the legislation was not surprising given that she had long supported the 1993 Religious Freedom Restoration Act, signed by her husband for the purpose, in Bill's words, "to protect a reasonable range of religious expression in public areas like schools and workplaces."[24] In addition, Hillary had written on this point in *It Takes a Village*, agreeing with conservative Christians on the importance of religion in schools. Quoting her husband, her book noted that "nothing in the First Amendment converts our public schools into religion-free zones, or requires all religious expression to be left behind at the schoolhouse door." She also cited these words from Bill: "[R]eligion is too important in our history and our heritage for us to keep it out of our schools."[25]

The Workplace Religious Freedom Act, which any reasonable person would support, had opponents among Senator Clinton's staunchest allies: Planned Parenthood and the National Women's Law Center foresaw intolerable instances of "anti-choice" injustice, such as a situation where a pro-life nurse might request to not provide the "morning-after" pill to a rape victim, or a Catholic pharmacist might as a matter of conscience refuse to dispense birth control. For these pro-choice organizations, religious freedom should not be permitted to overrule the unshakable right to an abortion. In many ways, it was surprising that Hillary did not agree with them; perhaps she did, but overlooked how her legislation would conflict with abortion rights.

It was this narrow opposition from activist groups that might have explained why, as the *Village Voice* put it, "[Mrs.] Clinton's office has been notably quiet about her involvement" in the bill.[26] The bill placed Senator Clinton in a conundrum, forcing her to choose between freedom of conscience and heresy in the church of abortion feminism.

Another Catholic Uproar

As Hillary struggled to connect her new language with new actions, the issue of abortion continued to hound her. Though she had proven herself willing to incorporate her own concept of faith and morality into the discussion, the issue seemed to be further separating her from the religious "values voters" she had hoped to attract. In addition, the issue was not going away: The entire cycle with Canisius four months earlier was about to repeat itself, this time at another Catholic college.

Just days after Senator Clinton voted to nullify the "Mexico City Policy"—by which President Bush in 2001, by executive order, stopped U.S. funds from supporting overseas "population assistance" programs that promoted abortion—Marymount Manhattan College in New York City announced that Mrs. Clinton would be delivering its commencement address on May 20 and would also be receiving an honorary doctoral degree.[27] Once again, Catholic groups were outraged and marshaled the troops.

Patrick Reilly, president of the Cardinal Newman Society, which establishes Newman Centers for Catholic students at colleges all over the country, said it would be a "scandal" for Marymount Manhattan to proceed. He made specific reference to Senator Clinton's vote against a ban on partial-birth abortion, her advocacy of embryonic stem cell research, and her declaration that contraception was "basic health care for women." Further, Reilly noted that Marymount Manhattan's actions defied the "Catholics in Political Life" document approved by the U.S. Conference of Catholic Bishops—the same guidelines that the bishop of Buffalo had underscored during the Canisius fiasco.[28]

Reilly wrote to Marymount Manhattan College president Judson Shaver, urging him to cancel Hillary's appearance "in order to restore fidelity to the college's Catholic mission, obedience to your bishop and public trust in your commitment to not lead astray the college's students, your employees and the general public." He likewise mailed a letter to New York's Edward Cardinal Egan seeking "immediate action to prevent scandal in the Archdiocese of New York." In that letter, Reilly spoke for those in the church who were fed up with the leftward drift of many Catholic colleges, some of which had moved so far from their moral and doctrinal underpinnings that it was difficult to call them Catholic. "After decades of scandal . . . the bishops drew a line in the sand," Reilly said. "No college that deliberately crosses that line deserves the label 'Catholic' or the support of the faithful—most especially monetary support."[29]

This was nothing new for the Catholic college founded by the Religious of the Sacred Heart of Mary in Tarrytown, New York; the previous year's commencement speaker was New York Attorney General Eliot Spitzer, another strong proponent of abortion rights. "Last year Spitzer, this year Hillary Clinton," Reilly remarked. "Can anyone trust this college's claim to a Catholic mission? They seem to have tried their hardest to honor the church's opponents in the fight against abortion, and now they publicly brush aside the bishops' clear expectations. We trust there will be consequences from the bishops, but lay Catholics also need to stand up and confront this scandal."[30]

What was worse for Reilly and others was that Marymount Manhattan boasted about its pro-choice commencement speaker the same month that Pope John Paul II had died: "Pope John Paul II has been laid to rest, but his legacy continues to be desecrated by heretics and public dissenters in 'religious studies' departments . . . and college officials who fail to uphold their colleges' Catholic mission," Reilly said.[31]

Joseph Starrs, director of the American Life League's Crusade for the Defense of Our Catholic Church, also affirmed that the invitation was "a clear violation of the stated policies of the U.S. Conference of Catholic Bishops." Starrs maintained, "By giving Senator Clinton an honorary degree, the college is thumbing its nose at the bishops and willfully disregarding a very clear mandate."[32]

Whether liberal Catholics liked it or not, Starrs and Reilly were accurate: The bishops had drawn a line, and the college in Manhattan was ignoring it. Senator Clinton was now in the middle of the crossfire. The college president clearly was unconcerned with that line, saying that he and the college were "honored and thrilled" that Senator Clinton would be addressing "the Marymount Manhattan community." "Sen. Clinton has worked diligently on behalf of New Yorkers and the nation, and with her record of service," wrote Shaver, "I have no doubt that her message will inspire our graduates and their guests."[33]

The negative press that Mrs. Clinton received was significant: That May, the Cardinal Newman Center monitored commencement speakers and honorees at all 220 Catholic colleges and universities in the United States, and created a list of those that were ignoring their charter and the wishes of the bishops. That list was widely published by Catholic Web sites and diocesan and national newspapers, including the front page of the *National Catholic Register*. Mrs. Clinton's name was the one most prominently featured at the top of those lists. Rather than succeeding in reaching out to what she rightly perceived as a crucial religious constituency, Mrs. Clinton had become a poster girl for Catholics outraged at Christian politicians who were unabashedly pro-choice.

Each of these occasions generated large volumes of unwanted press for Hillary on Web sites and in newspapers where she had badly wanted to make advances, and where it was pointed out repeatedly that while Hillary was a self-described Christian, she was also unacceptably "pro-abortion," as these sources labeled her—a term Mrs. Clinton detests.[34] Sure, the mainstream press tried to ignore these controversies that consumed pro-life Catholics, but that was not the case, however, with those constituencies that Mrs. Clinton hoped to bring along.

Just then, she handed them another issue for outrage: She opposed a Republican bill that would make it illegal for anyone to help an underage girl go out of state to have an abortion without her parents' consent.[35] Then, three months later, in August 2005, Senator Clinton ripped the nation's top health official—Health and Human Services

Secretary Michael Leavitt—for a "breach of faith" over the FDA's failure to decide whether to let women buy emergency contraceptive pills without prescriptions.[36]

In all, Marymount, and Canisius before it, was a wake-up call, demonstrating to Mrs. Clinton that language alone would not placate the Catholic constituents she was trying to win. The only way to win, it seemed, was to change.

The Supreme Court Opens

Those pro-life Catholics whose skepticism over Hillary resulted in the furors at Marymount and Canisius soon found vindication in her attempts to block two pro-life Catholic U.S. Supreme Court nominees who could begin turning the tide against *Roe v. Wade*.

The debate over the role that abortion would play in Supreme Court vacancies was nothing new, and in fact this had been a decisive issue in her 2000 Senate campaign. In a debate with Rick Lazio, when asked about the kind of U.S. Supreme Court nominee she would support, Mrs. Clinton had made clear that the justice's position on abortion would be the single most important factor, insisting on a so-called litmus test. "I think the fate of the Supreme Court hangs in the balance," she said. "If we take Governor Bush at his word, his two favorite justices are [Antonin] Scalia and [Clarence] Thomas, both of whom are committed to overturning *Roe v. Wade*, ending a woman's right to choose. I could not go along with that." In the next sentence, she issued her code language for abortion rights in the years ahead as each pro-life pick by President Bush was nominated for the high court: "In the Senate, I will be looking very carefully at the constitutional views as to what that nominee believes about basic, fundamental, constitutional rights."[37]

Standing by this language, she used similar words to oppose the appointment of John Roberts to the court and in her statement on the death of Chief Justice William Rehnquist. In her September 5, 2005, statement on Rehnquist's passing, she urged he be replaced by a judge who places "fairness and justice before ideology." She repeated these words two weeks later in a September 22 press release rejecting Bush's promotion of Roberts to fill Rehnquist's seat; the vast majority of the two-page press release dealt with *Roe v. Wade*.

Only a month later came the next opening on the high court, and on October 31, 2005, Mrs. Clinton issued a statement saying that Bush's pick of Judge Samuel Alito "raises serious questions about whether he will be steadfast in protecting our most fundamental rights," meaning abortion rights. As with Roberts, she said she wanted to examine Alito's record carefully to ensure that he places "fairness and justice"—additional code words for abortion rights—"before ideology."

In a statement from her office on January 27, 2006, Senator Clinton said that confirming Alito would halt "American progress" and "the ever-expanding circle of freedom and opportunity." Judge Alito, said the senator, "would narrow that circle." This statement on Alito was indicative of how her conciliatory language was more linguistic flourish than evidence of a genuine shift on abortion policy.

Indeed, if 2005 saw her extend a figurative olive branch to the right and middle, 2006 saw her seemingly retract it, as she went back out on the campaign trail and back into the churches. On January 16, Mrs. Clinton delivered a speech to an African American congregation at the Canaan Baptist Church of Christ in Harlem. It was Martin Luther King Jr. Day, a time of national remembrance and reconciliation on the divisive, hurtful history of race relations. Yet what Mrs. Clinton had to say that day was blistering. The junior senator from New York told the black audience that Republican congressional leaders had been running the U.S. House of Representatives "like a plantation." She added, in a slap that was tame by comparison, that the Bush administration would go down as "one of the worst" in U.S. history.[38]

Suddenly Hillary was back to her rabble-rousing self, yelling at the crowd as a host of other white New York Democrats up for reelection sat nearby and watched: "When you look at the way the House of Representatives has been run, it has been run like a plantation, and you know what I'm talking about!"[39]

Understandably, the congressional leaders at the receiving end of these remarks were offended. "I've never run a plantation before," said the Speaker of the House, Dennis Hastert (R-Ill.), who, by Mrs. Clinton's implication, was the equivalent of a nineteenth-century Southern slave owner. "I'm not even sure of what kind of association she's trying to make. . . . I think that's unfortunate, but I'm not going to comment any further."[40]

One of Hastert's colleagues, liberal Republican Peter King (R-N.Y.), a close friend of both Clintons, who had defended Bill in the impeachment proceedings, for which he was trashed by many in his own party, said of Mrs. Clinton's remarks: "It's definitely using the race card. It definitely has racist connotations. She knows it. She knew the audience. She knew what she was trying to say, and it was wrong. And she should be ashamed."[41]

She had defenders, including the host of the event, racial activist Reverend Al Sharpton. "I absolutely defend her saying it," said Sharpton, adding that Senator Clinton was merely reiterating what he himself had said. Likewise, Mrs. Clinton's spokesman defended the remarks, saying she was simply trying to underscore the point that the GOP House leadership stifled substantive debate—a point, it seems, that could have been made in a different way, on a different day, and in a different place.[42]

If the former first lady was looking to win middle-of-the-road voters, this was not the way to do it. Either she had spoken emotionally and loosely, or, worse, she had deliberately employed very divisive tactics—in a house of God, no less, and on Martin Luther King Jr. Day—merely to rile up a political base that she needed in the fall.

More rancorous statements followed from Mrs. Clinton throughout 2006, belying her strategy of a new tone aimed at moderate voters, and instead suggesting that her campaign felt a need to feed red meat to the hard-left base.

In March 2006, the senator employed the Bible to slap Republicans who backed a new immigration bill that she claimed was not only un-American but un-Christian as well. "It is certainly not in keeping with my understanding of the Scriptures," Clinton judged. "This bill would literally criminalize the Good Samaritan—and probably even Jesus himself."[43] Mrs. Clinton must have known that George W. Bush considers the story of the Good Samaritan one of his favorite parables, and she was striking at the core of what religiously motivates the current Republican president.[44]

Here again, she clashed with her good friend Representative Peter King (R-N.Y.), who cosponsored the bill. "I don't think Jesus would have defended alien-smuggling gangs," King fired back at the junior senator from New York. "I don't think Jesus would favor hundreds of immigrants dying in the desert."[45]

Notably, the Catholic Church was sympathetic, fearing, as did Mrs. Clinton, that the immigration measure would turn a priest who counsels an illegal alien into a criminal. At last she had found an area of agreement with Rome.

One of the more surprising of these remarks came during a May 2006 assertion that right-wing "ideologues" were to blame for abortion. According to Mrs. Clinton, by denying women access to contraceptives—here she must have had in mind Catholics in particular—these right-wingers left women with no choice but to end their unwanted pregnancies with abortions. This movement to withhold contraceptives, said the senator, in strong language, "was started by a small group of extreme ideologues who claim the right to impose their personal beliefs on the overwhelming majority of the American people." She added: "They're waging this silent war on contraception by using the power of the White House and their right-wing allies in Congress"—"and so far, they're getting away with it."

Mrs. Clinton seemed to be identifying a new right-wing conspiracy, this time against contraception. Her assertion harked back to the 2004 NARAL address, which, her speeches in 2005 had implied, was a thing of the past. Now, however, she was again granting extraordinary powers to nefarious, conspiratorial "anti-choice" forces who, she suggested, had somehow in silence been able to impose their views on contraception upon the majority of America—a country in which contraception had long been legalized and was under no threat of illegalization. After all, most devout Protestants, and even millions of Catholics, use contraception.

She turned the matter, as she had before, into a class issue, saying that "low-income women, denied access to contraception, are having more unwanted pregnancies—four times as many as those for higher income women. And almost half of all unwanted pregnancies end in abortion."[46] This may have been a clever tactic for dealing with pro-lifers like Alveda King, niece of the late Martin Luther King Jr., who have been adamant in blaming pro-choicers for unintended "genocide" against unborn babies from low-income African-American communities through legalized abortion, an argument Mrs. Clinton has heard. In fact, the Reverend Jesse Jackson, Hillary's old prayer partner, used to level this charge back when he was pro-life, before he began pursuing the Democratic presidential nomination. In essence, Mrs. Clinton was reversing the argument onto pro-lifers, or, at least, onto "right-wingers," by contending that they are in large part to blame for any resulting deaths by denying contraception to low-income women.

Stem Cells

In 2006, the life debate took on new fronts as the embryonic stem cell research (ESCR) discussion made its way to the halls of Congress. ESCR advocates the creation of human embryos for the purpose of scientific research. Embryos used for such purposes cannot, of course, reach their full potential to become human life; they are dissected, with the unneeded components discarded.

Proponents argue that ESCR holds great potential for the cure of certain diseases, from Alzheimer's to Parkinson's—or, as Mrs. Clinton put it, "countless" diseases. High-profile advocates include former first lady Nancy Reagan and the late actor Christopher Reeve. Opponents counter that this potential is greatly exaggerated. "Embryonic stem cells are not going to be the source of a cure for Alzheimer's," says Dr. James Dobson, a vocal foe of Mrs. Clinton on this issue.[47] Moreover, opponents fear that ESCR is the first step on the road to a Brave New World of "harvesting" or "farming" fetuses strictly for the purposes of taking their "spare parts." They maintain that these human embryos are human life at its earliest stage of development, raised and then snuffed out for the benefit of the living.

Thus, ESCR is an obvious pro-life issue for pro-lifers, but it seemed like less of a pro-choice issue to pro-choicers, since it does not involve abortion. Nonetheless, pro-choice advocates like Mrs. Clinton, for various reasons stemming from the politics of the abortion debate and general "life vs. choice" philosophies, support ESCR.

Although she ended up once again in opposition to the right-to-life movement on this subject, too, Senator Clinton tried to massage her rhetoric while not changing her position. An excellent example was her cosponsorship of S. 471, the Stem Cell Research Enhancement Act, which would have allowed for federal funding for research on new stem cell lines extracted from human embryos.

The July 18, 2006, statement from Senator Clinton's office, which included the full text of her remarks on the Senate floor calling for passage of the legislation, was remarkable in that it never once used the words "embryo" or "embryonic" among its thirteen hundred words. Mrs. Clinton spoke of how "stem cell research" holds the promise of "new cures and treatments for countless diseases and millions of Americans with chronic and curable conditions." For example, she said, "The wide range of applications for stem cells may lead to unparalleled achievements on behalf of research concerning Alzheimer's disease," as well as spinal cord injuries, diabetes, Parkinson's, Lou Gehrig's disease, and a host of others she listed. She four times mentioned her "dear friend" Christopher Reeve, who was confined to a wheelchair and who eventually died from a spinal cord injury, as well as referencing Reeve's wife and son, plus more than once mentioning "children" and even "grandchildren," and words like "suffering."

Only a well-informed observer already familiar with the legislation would know from reading the senator's statement that it concerned *embryonic* stem cells. It is not unfair to describe the statement as deliberately deceptive for merely mentioning "stem cells," because no one opposes research on *adult* stem cells, or stem cells acquired through umbilical cord blood or bone marrow—the debate is over stem cells acquired through the killing of an embryo. Senator Clinton's statement did not make that crucial distinction.

The senator's statement also revealed a change in tone: Mrs. Clinton ended her plea on the Senate floor by affirming her "respect" for those who disagree with her on the issue. She even called for erecting a clear "ethical fence around this research," with "very strong prohibitions and penalties for people who don't pursue the research in the way that we set forth." Of course, opponents of ESCR believe that the research itself is unethical, and that the ethical breach comes from creating and discarding the embryos simply for the sake of their material in the first place. Nonetheless, despite the respectful rhetoric, Senator Clinton again found herself deeply at odds with many Christians, and yet again with the Catholic Church.

The Republican-controlled Senate passed the bill, with nearly every Democrat voting in support and enough liberal Republicans crossing over. The next day, on July 19, 2006, President George W. Bush exercised the first veto of his presidency. Holding a press conference surrounded by eighteen families with children who had once been frozen embryos, Bush said that if the Stem Cell Research Enhancement Act had become law, "for the first time in our history we would have been forced to fund the deliberate destruction of human embryos, and I'm not going to allow it. . . . [W]e all began our lives as a small collection of cells." Bush said that Americans "must never abandon our fundamental moral principles in our zeal for new treatments and cures."

Despite the veto, the fact that the bill had passed in the Senate showed that some of the tides were turning. Hillary's words had played a role in shaping the Democratic victories of the year, and it seemed that she was on the cusp of something bigger. Mrs. Clinton's remarks during 2006 had been quite bold. Were these statements of election year politics, or were they Hillary's true colors? Largely gone was the conciliatory moral and religious tone from the previous year; in its place were heated, confrontational words that proved to the Democratic faithful that she was a voice that could be trusted. If the goal was to distance herself from the religious right and excite her liberal base for the purpose of winning over New Yorkers again in November 2006, she was achieving it in spades.

CHAPTER 15

Hillary and the Faith Factor

Since the 2006 midterm elections, Hillary has focused her attention on a different goal, one that while not new, appears to be attainable. The next step in the life of the girl from Park Ridge is a giant one: She has a serious shot at becoming the first woman president of the United States. To win the presidency in 2008, Hillary must not merely prove herself a centrist, religious Democrat, but needs to overcome past negative perceptions on the religiousness of her and husband. A 2000 poll by the *Wall Street Journal*/NBC News found that only 12 percent of the public described Hillary as "extremely/very religious" and 25 percent found her "not that religious"; of all the figures polled, only her husband scored lower.[1] The poll found that the American public did not perceive the Clintons as religious people, almost certainly an extension of the fact that they did not see Bill's private behavior as reflective of a man possessing and motivated by Christian ethics and values.

Hillary not only wants to reverse this perception but also wants to avoid the character charges that plagued her husband through his final days in office. Indeed, it was telling that as George W. Bush was being inaugurated in January 2001, an authority as respected as *Time*'s dean of presidential correspondents, the reserved Hugh Sidey, raised eyebrows on CNN by calling Mrs. Clinton's husband a "disgrace" who lacked dignity, while over at MSNBC, a veteran Democrat like Pat Caddell judged Bill Clinton "white trash" who had "no class" and had "slimed, slimed, slimed" the presidency.[2]

For Bill, the character issue was so bad that it overcame any ability by him to ever successfully position himself as a religious man, and as a result, Hillary suffered from guilt by association. Mrs. Clinton got slimed herself during that presidency. Conservative columnists like George Will lumped her into his Clinton "vulgarian" category, and William Safire deemed her a "congenital liar." From the left, Maureen Dowd served up a litany of irreverent descriptions of the former first lady.[3]

Yet Hillary now suddenly seems beyond that, almost as if the 1990s had never happened. To go back and read through clippings from the previous decade is to revisit a world that no longer exists—and very much to Hillary's benefit.

This was evident in a May 2005 *USA Today*/CNN/Gallup poll that found a majority of Americans "likely" to vote for her for president, and with her pulling higher "moderate" ratings than ever, as 54 percent judged her a liberal but 30 percent a moderate. Said Andrew Kohut of the Pew Research Center, "This may . . . reflect that she has been recasting her image as a more moderate person." She commanded nearly as much support as Governor George W. Bush did in 1998, two years before the 2000 election. On the other hand, the poll noted a continuing problem for Mrs. Clinton: an abnormally high percentage of voters that strongly dislike her, especially men.[4]

Nonetheless, as Kohut noted, "over time, Clinton fatigue has dissipated." People are not looking back at the Clinton years with many of the same attitudes. Thinking has shifted so much that Ed Klein, author of a book highly critical of the former first lady, now says, "Hillary acts as though she has been chosen by God"—she has "this aura about her as if she has been chosen by God to lead us."[5] Klein himself reflects the polarization over Hillary, as he says, presumably metaphorically, that both of the Clintons "have sold their souls to the devil in order to achieve power."[6]

Whether Hillary feels that she has been chosen by God is something she has not shared openly; she certainly has never made such a bold declaration publicly. Regardless of whether God has chosen her, the immediate task will be for her to convince millions of Americans to choose her in November 2008. This will be a difficult task, as her strategy to position herself in 2008 as a religious moderate is constantly fluctuating between her liberal constituents on the left and the Christian center of the country that she tries to appease.

In her attempt to win over the American public, Hillary will no doubt bring her religious coattails to the forefront of the discussion. However, one of the sizable questions about her faith in this election cycle will be how she frames her opinions in the primary election so that some of her more religious ideas appear palatable to Democratic voters. Despite her efforts in the Senate to convince voters of her faith and moderation, few Democratic primary voters are likely to use these issues as their bellwether for their presidential candidate.

As such, Hillary will certainly rely on her more impassioned liberal credentials to convince primary voters that she agrees with them on the issues that matter to Democratic voters, such as abortion, gay marriage, the war in Iraq, health care, the environment. Of less importance will be the role that her faith plays in her policies, and how faith shapes her worldview. While any candidate in a modern campaign must watch his or her verbiage closely, Hillary in particular must be aware of the accusations that follow her, especially those that charge her with hypocrisy on spiritual issues. As someone who seems to tailor her spiritual opinions to the crowd she is addressing, she must exercise great caution to avoid having some of her more moderate (or more liberal) views thrown back at her during her quest for the nomination.

If she succeeds in making it out of the primary, she will find herself facing a much different set of obstacles. Suddenly she will need to reinforce her religious and moral pedigree, emphasizing her religion in a way that will force all Americans to identify with her as a God-fearing woman. She must, for example, remind Americans of the bill she cosponsored with Rick Santorum to promote the acceptance of faith in the public arena, a bill that can be a big winner for Hillary with cultural conservatives.

Similarly, Hillary might be expected to roll out a comprehensive plan of faith-based initiatives, so that religious Americans can see that her words about the social justice implications of Jesus are more than just words; they are ideas that she is willing to back up with action. Coming out in clear support of faith-based programs with new goals of her own could provide the boost that she is looking for with many faith-oriented Americans.

Though she is on the record as consistently opposing gay marriage for historical, cultural, and biblical reasons, this seems like another topic of increasing ambivalence by Senator Clinton that would need to be addressed in any campaign. It has caused problems for her politically, and has caused flip-flops in how she deals with competing New York constituencies. The subject, for her, comes up not only each time it is raised in Congress, but annually in each St. Patrick's Day parade in New York City. As of the writing of this book, the latest example, in March 2006, witnessed Senator Clinton's decision to march in the parade.[7]

She marched in the parade when she was running for the Senate seat in 2000, did not in 2001—saying she had a schedule conflict—sat out again in 2002, 2003, 2004, and 2005, and then marched again in 2006, another election year. Thus, of her six years in the Senate, she marched in the two years in which she was campaigning for election and sat out in the years she was not running. Predicts Bill Donohue of the New York-based Catholic League: "No doubt she'll march again for the next couple of years because she wants to be elected president in 2008." Adds Donahue: "Hillary Clinton's chameleonic response to the St. Patrick's Day Parade is grounded in politics and deceit. She will pander to gays one year, to Irish Catholics the next. She's fooling no one save her loyalists."[8] Actually, the loyalists are not fooled, and are peeved—but will vote for her.

Regardless of the externals, she seems committed to her apparently core position that gay marriage is not acceptable. Yet she appears to buckle in her willingness to be frank and bold on this issue, depending upon whom she is talking to, and opening herself to the moniker of "Slippery Hillary," the heir to her husband's "Slick Willie."

Nevertheless, if she is going to win religious voters, she will need to come clean on this stance and tell America that though she supports gay rights, she does not support same-sex marriage. One of the few areas where Hillary has been publicly consistent is on this subject: She has never said that she supports gay marriage. Assuming that she stays a decisive clear voice on the issue, this is a useful stance that could offer a significant boost to her attempt to appear moderate as she runs for president.

Programs and legislation in areas like this could go a long way to revealing the forcefulness of her faith, and she will certainly do her best to supplement them with a lot of campaign trail talk that demonstrates her level of comfort with God and the Bible. The goal in these entreaties is not solely to win people over with her perspective, but to convince voters that her personal theology is rooted in more than mere politics and make them believe that she would be a president who turns to God to help her lead the country.

But even if she succeeds in crafting a set of policies that caters to the religious center, Hillary is likely to still find herself facing an uphill battle with other so-called values voters. Because of both the lingering stigma of her husband's presidency and her refusal to cede ground on controversial religious-moral issues such as abortion, Hillary will struggle to win the hearts and minds of the country's religious middle.

As Christian voters—both Catholic and Protestant—look to Hillary's past, present, and future, they will see a candidate whose religious life is evident, but whose religious politics are in turmoil. One such issue has to do with the entertainment culture. Hillary Clinton is often described as a "prude" by advisers close to her and her husband, and as such, it has long been rumored that she is uncomfortable with the licentious, violent culture depicted by the Hollywood mavens that supported her husband's presidential bids. Hugh's daughter would be the first to say that Hollywood should not be proud of the sewer pipe of filth it has channeled into the hearts and minds of young people.

Despite her personal preferences, a difficulty is that while the values of Hollywood might clash with her own, much of Hollywood has contributed to the Clinton war chest. In addition to these donations, there may be a belief in the pornography industry that having the Clintons back in the White House could be smooth sailing for porn. Indeed, a scarcely acknowledged fact concerning Bill Clinton's presidency was the boom in the porn industry. "[W]hen Clinton comes in," said Mark Cromer, producer of the X-rated Hustler Video, "it's definitely blue skies and green lights and fat bank accounts."[9]

The green light is traceable directly to the policies of the Clinton administration. In the 1980s, the porn industry had been on the ropes, targeted by the Reagan administration, notably the vigilant efforts of Attorney General Ed Meese. Then, in the early 1990s, the arrival of the Clinton administration forced federal prosecutors to halt many investigations since there were "different priorities" under the Clinton–Janet Reno Justice Department. "Under Attorney General Reno," noted a special investigative report by PBS's *Frontline*, "federal prosecutions slowed dramatically, and the obscenity task force effectively went out of business."[10]

Hustler magazine, for instance, was so ecstatic with the Clinton era and all the happenings, legally and symbolically, that founder and porn king Larry Flynt went to bat on behalf of Bill Clinton during the Lewinsky scandal, investigating the sexual lives of the president's opponents. So busy was Flynt that Michael Kelly in the *Washington Post* christened him "the president's pornographer."[11]

Thus when combined with the advent of Internet pornography, the porn industry was resurrected during the 1990s, a development that surely did not please Hillary, since the industry is degrading to women. Now, under the Bush administration, the halcyon days of the 1990s are gone. After his confirmation, Bush's Attorney General John Ashcroft immediately met with anti-porn prosecutors and advocates in an attempt to avoid a continuation of Clinton administration policies.[12]

While these policies were her husband's and not her own, the close Clinton ties to the entertainment industry leave her vulnerable to criticism as she pursues a family values platform. It will be hard to find a solid bloc of Christian voters willing to choose Hillary if she is unable to appear convincing about the values our culture espouses through entertainment.

Yet another point that she must face during her appeal to religious Christians is the role that stem cell research—embryonic and non-embryonic—should play in the future designs of medical technology. This issue and the disagreement over its medical potential is shaping up to be one of the great public and scientific debates of our time.

Although Hillary lately has been gracious in acknowledging the opinions of the other side in this debate, she has not altered her stance in a way that will mollify detractors. In the bills that have come in front of the Senate, Hillary has voted in favor of embryonic research, voting "yea" on, among others: S. 1557, the Respect for Life Pluripotent Stem Cell Act; S. 876, the Brownback-Landrieu Human Cloning Prohibition Act; and S. 1520, which also looked to promote cloning of human embryos.

As the Missouri Senate race in 2006 demonstrated, embryonic stem cell research is quietly shaping up to be an issue that people in this country—values voters and secular voters alike—are willing to make a centerpiece of their voting ideology. With Hillary in the White House, the American public will know that bills favoring embryonic research will start to pass, and for many Christian moderates, this will be an important consideration when casting their ballot in November 2008.

Despite the emerging strength of the pro– and anti–embryonic stem cell movements, as this book has made clear, there is no religious issue that holds greater risk for Hillary's position as a religious Democrat than abortion. Not only has she been steadfast in her support of a woman's right to an abortion, but a careful reading of her speeches on the subject reveals the development of a kind of abortion apologetics by Hillary, where she has not merely affirmed a woman's right to abortion but has generated a series of counterattacks against various arguments that the pro-life movement uses to chastise abortion rights advocates. On this issue, Hillary Rodham Clinton the Methodist is a Fundamentalist, as dogmatic and filled with conviction as the most fire-and-brimstone preacher. With the exception of her talk in Albany in January 2005, she has spoken of abortion rights in an absolutist way, pitting good vs. evil, at times demonizing the other side.

Noticeably, she makes no mention of her Methodism when discussing the subject. Of course, if she did, it would not change anything: The United Methodist Church's *Book of Discipline* has been a major source of guidance for her on moral questions; the degree to which she matches her church on social concerns is uncanny.[13] And her church's position on abortion is not pro-life; to the contrary, the United Methodist Church supports legalized abortion and is even a member of the Religious Coalition for Reproductive Choice.[14] The United Methodist Church's official statements on abortion have both reinforced and helped guide Hillary into her pro-choice position.[15]

Within her position, Mrs. Clinton has carefully avoided stating whether she believes life begins at conception, a sign of her shrewdness and a smart display of understanding that if she made such an acknowledgment, her pro-choice position drifts from its moral moorings. As noted by Greg Koukl of the Los Angeles–based ministry Stand to Reason, if the object in the womb is not a life begun at conception, then whatever one chooses to do with it is of no concern. The pro-choice argument is uncontroversial and unimpeachable if the object in the womb is not a human life. But once one acknowledges that the object is human life, moral considerations change. Hillary understands the stakes of conceding that life begins at conception. (The Democratic Party's last presidential candidate, John F. Kerry, did not know any better.) Surely this life-begins-at-conception paradox has occurred to Mrs. Clinton and thus likely troubles her, which is why she has avoided acknowledging it. That said, there is reason to suspect, however, that she personally believes that life does begin at conception. Consider two sources:

In an important piece in *New York* magazine, which led with Hillary's Albany speech, reporter Jennifer Senior shared a telling anecdote. Senior noted that the day after the speech, Clinton intimate Harold Ickes was in Washington expanding on Hillary's remarks; he was doing some legwork with Senator Clinton's pro-choice constituency: "I'm sorry, but when push comes to f—ing shove," waxed Ickes, "my belief is that life begins at conception. And I think Hillary understands how hot-button this issue is for Democrats."[16] Presumably Ickes was explaining, inelegantly, that many Democrats, when pushed, concede the conception point—possibly even Hillary herself.

Of course, that is an inference. Less of an inference on her thinking is this assessment by her husband in his 2004 memoirs: "Everyone knows life begins biologically at conception," says Bill Clinton, the man in whom Hillary has confided and with whom she has had more discussions than with anyone else on more subjects, including abortion. One would think that "everyone" must include Hillary. Bill added, two sentences later, "Most people who are pro-choice understand that abortions terminate potential life."[17]

Bill Clinton continued the thought, taking it to a theological level: "No one knows when biology turns into humanity or, for the religious, when the soul enters the body."[18] Mother Teresa had an opinion on the question, telling the Clintons during the National Prayer Breakfast: "But what does God say to us? He says: 'Even if a mother could forget her child, I will not forget you. I have carved you in the palm of my hand.'"

Pro-lifers are frustrated by what they see as Mrs. Clinton's cognitive dissonance on this matter: They find it odd that she champions the rights of children but only once those same children are out of the womb. She literally sees a world of difference between a child born on, say, March 23, and that same child on March 22 or February 22 or September 22; the former must be legally protected, whereas the latter's life is at the discretion of the mother alone. She makes statements like this: "The very core of what I believe is this concept of individual worth, which I think flows from all of us being creatures of God and being imbued with a spirit."[19] Presumably, she judges that the March 23 child is imbued with a spirit but not the same child days or weeks or months prior to birth.

Because of her longstanding support for abortion rights, there is little if anything that Hillary can do to have credibility in her assertion that she wants to find "common ground" with pro-lifers and to make abortions safe, legal, and *rare*. The only way that she might be able to accomplish convincing them of this would be for her (by 2008) to find and vote for a bill that does just that. At this point, however, it is difficult to imagine any such bill.

In an analysis of her abortion views, the *New York Times* noted that the only part of the abortion debate in which Hillary seems to have shown a detectable change over the years is parental notification—to which she has moved ever further to the extreme. The *Times* stated that while in Arkansas, Hillary supported laws to notify parents when a minor sought an abortion, unless an exception was granted by a judge. Now, in New York, reported the *Times*, she supports the state law of "informed consent," which is very different, and much less restrictive. Under informed consent, health care providers are told to give information about the risks of an abortion as well as other medical options only to patients, not to parents.[20]

At the writing of this book, Hillary had not switched her position on a single aspect of abortion or embryonic stem cell research. She opposed the following Senate bills, all supported by pro-life and rejected by pro-choice groups: S. 755, Informed Choice Act; S. 511, RU-486 Suspension and Review Act; S. 403, Child Custody Protection Act (parental notification); and S. 51, Unborn Child Pain Awareness Act. Among them, her opposition to the Informed Choice Act had pro-lifers fuming as it sought to provide aid for ultrasound machines, so pregnant women considering an abortion would be able to view the unborn child in their womb before making a final decision. The abortion industry opposes these machines for the reasons that pro-lifers embrace them: Data show that when viewing the humanity of their unborn child through an ultrasound image, the vast majority of women considering an abortion change their minds. Thus, conservative Christian organizations like Focus on the Family have dedicated entire ministries to raising millions of dollars for the sole purpose of purchasing ultrasound machines for crisis pregnancy centers.

Titling the act the Informed Choice Act was clever, since the goal of ultrasound technology is to give the mother maximum information about what is growing inside her—so much information that she can view the moving life inside. This allows for a fully "informed" "choice." And yet Hillary, even though she states emphatically that her goal is to make abortion rare, does not support the act. Politically, this would have been an easy bill for her to back; she could do so by arguing that it indeed more fully informed a woman's choice. The abortion lobby would have protested but would surely still view her as a reliable stalwart for the cause. By opposing it, pro-lifers will claim Hillary is "pro-abortion" more than "pro-choice"—a charge that really gets under her skin.

Also in keeping with providing would-be mothers with more information about the choice of a lifetime is the Unborn Child Pain Awareness Act, which mandates guidelines requiring an abortion provider to inform pregnant women considering an abortion that the child that they are about to abort feels pain during the abortive process, another determinative piece of information that often has the effect of changing the mother's mind. Here, too, Mrs. Clinton voted against the bill.

It is very telling how Mrs. Clinton mirrors her husband on this issue. Her husband held a moderate position in several policy areas, but not abortion. Bill Clinton was the "Planned Parenthood President," the most pro-choice individual ever to occupy the Oval Office. While he constantly and consistently searched out areas on which he could stake a moderate position—such as welfare reform—his stance on abortion was never middle-of-the-road and always extreme.

Significantly, recent memos obtained through Freedom of Information Act requests by the conservative legal watchdog Judicial Watch revealed what many long suspected: that Mrs. Clinton herself was behind Bill's unwillingness to move closer to the middle on the abortion issue. Among the documents was a memo from Domestic Policy Council staffer Bill Galston, written only months into the start of the Clinton presidency, suggesting that President Clinton "lower the public profile" of the increasingly radical abortion policies of the administration. Galston cited a number of abortion topics and strategies, including those of the abortion "working group" that included "the First Lady's office." At the "decision" section of the Galston memo, Bill Clinton handwrote a note asking, "What does Hillary think?" This and other memos led Judicial Watch to justifiably conclude that Mrs. Clinton was "at the center" of Clinton administration decision making on the issue—and was even "the driving force of the White House's abortion policy." Thus, Hillary is likely the primary reason that Bill refused to moderate on this one issue.[21]

Not coincidentally, as an elected figure herself, Mrs. Clinton has followed this same path. Hillary likewise looks hard for issues on which she can stake a moderate stance, and she has found a few, namely in defense and foreign policy. Like her husband, however, she refuses to moderate on the subject of abortion. A President Hillary Clinton would recapture the crown from her husband, taking abortion rights to new levels.

What exactly would she do? In addition to filling the federal bench with pro-choice judges, she would reverse the following moves by President George W. Bush:

On his first day in office, Bush authorized a ban on all U.S. funding of international abortion rights groups, reversing President Bill Clinton's executive order. Even before that, when he was president-elect, Bush held a private talk with the pro-choice Republican Colin Powell, several weeks before naming him secretary of state. He told Powell that as his secretary of state Powell would be expected to purge any vestiges of the Clinton State Department's program to promote global abortion rights. Powell told Bush he would follow his lead.[22] For Bush, other Clinton reversals soon followed. In January 2003, Bush signed the "Sanctity of Life" bill, and then, two months later, in March 2003, he did not veto the Republican Senate's passage of a ban on partial-birth abortion. Several months later, in November 2003, he signed the ban passed by Congress. George W. Bush also began commemorating each January 22 anniversary of *Roe v. Wade* as a National Sanctity and Dignity of Human Life Day.

Of course, while pro-lifers dread the prospect of a President Hillary Clinton reversing these actions, the thought excites pro-choicers. Asked if he would expect Hillary to change these policies, William F. Harrison, the Arkansas abortion doctor and Hillary's personal friend and onetime ob-gyn, exclaims: "Oh, absolutely. . . . I hope to God she does." He and other pro-choicers are counting on Hillary: Though into his seventies, Harrison does not want to slow his rate of activity at his Fayetteville Women's Clinic; he plans to continue to perform about twelve hundred abortions per year. The key, says Harrison, will be whether the electorate can appreciate both the Clintons, whom he says history will judge "with a much more reasoned and rational mind than the idiots who have hated [them], seemingly for no more reason than Christ was hated."[23]

Seeing the clear rift between her stated policy stances and religious voters, Hillary has decided to try a new strategy: At the end of 2006, she hired Burns Strider, a leading party strategist on advising candidates how to reach out to pro-life evangelicals, a group that provided a surprisingly high number of votes in November 2006 to the pro-choice governor of Michigan, Jennifer Granholm, and the Ohio governor-elect, Ted Strickland.[24]

The hiring of Strider may be a signal that Mrs. Clinton realizes that symbolic gestures on the abortion debate, such as telling pro-choice crowds in Albany that pro-lifers are not cavemen, will not be enough to satisfy religious "values voters." The selection of Strider also demonstrates that Hillary recognizes that the Democrats' success in November 2006 took place during an unusual election that was more of a statement on George W. Bush's policies in Iraq than social issues like abortion. In 2008, Bush will not be on the ticket. Who will the Republicans run?

Regardless of the GOP nominee, even possibly a pro-choicer, Hillary will struggle to peel off pro-life religious voters until she takes concrete policy stances that soften her abortion extremism. Until that happens, she, like John Kerry and Al Gore, may find it difficult to change that blue-red state configuration in 2008. To the extent that she hopes that her faith will be a draw to some voters, and to the extent that she hopes to draw a single religious voter beyond the religious left, Hillary's position on abortion will be the determining factor.

Liberals may want to dismiss the abortion issue as a mere concern of right-wing "fundamentalists" who watch *The 700 Club*. Yet this issue extends way beyond Pat Robertson Republicans. It goes to Reagan Democrats, to blue-collar Democrats, to Blue Dog Democrats, to Southern Democrats, to Catholic Democrats, to Pennsylvania and Ohio and Iowa and Michigan Democrats and those in other traditional swing states.

One religious Democrat—perhaps the country's foremost—with his pulse on the issue is former President Jimmy Carter. On November 3, 2005, Carter was in Washington to promote his book *Our Enduring Values*. He spoke to reporters over breakfast at the Ritz-Carlton Hotel. "I never have felt that any abortion should be committed," said the former Democratic president, to the shock and dismay of the reporters. "I think each abortion is the result of a series of errors."[25]

Hillary might have even considered going that far herself, though she never would have dared to say what Carter said next: "I've never been convinced, if you let me inject my Christianity into it, that Jesus Christ would approve abortion." Carter went on: "I have always thought that it was not in the mainstream of the American public to be extremely liberal on many issues. I think our party's leaders—some of them—are overemphasizing the abortion issue." He surely had Senator Clinton in mind.

There are recent data to back this. A March 2006 Zogby poll further illustrated what has been evident for a long time: Hillary Clinton's abortion stridency may be a smash on Broadway, but it does not play in Peoria; it works for her in blue states like New York, but is a loser in red states like Nebraska—and she needs red states to be elected president. The Zogby poll found Americans strongly disagreeing with Mrs. Clinton on abortion matters like parental notification (by ratios of two to one for girls under eighteen, and three to one for girls under sixteen), federal funding for abortions abroad (more than three to one), and much more. By a ratio of two to one (59 percent to 29 percent), Americans believe that "abortion ends a human life." The poll, commissioned by Associated Television News President Brad O'Leary, prompted O'Leary to remark that the results "spell disaster for Democrats who try to run on the abortion issue. The abortion issue is this year's 'third rail' for congressional Democrats and for Hillary Clinton in 2008."[26]

But while abortion is a significant religious issue, it is not the only religious issue. A barometer of what Mrs. Clinton can expect to confront as she pursues the nation's highest office as a pro-choice Christian is the experience of the rising Democratic Party star who holds the junior Senate seat in Mrs. Clinton's home state, one whom she respects very much and who initially excited her greatly: Barack Obama.

Before *The Audacity of Hope*, Obama's best-selling book in late 2006, and before his sudden emergence as a near neck-and-neck rival with Hillary in early presidential polls, Obama gave an important speech on religion, a prelude to the freshman senator's surge in the months and year ahead. Specifically, in a June 28, 2006, address to the annual gathering of the Call to Renewal convention in Washington, Obama made a heartfelt appeal to religious voters on behalf of liberal Christian politicians.[27] E. J. Dionne, a *Washington Post* columnist with a special interest in religion and the politics of the Democratic Party, claimed that Obama's "eloquent faith" might have produced "the most important pronouncement by a Democrat on faith and politics since John F. Kennedy's Houston speech in 1960 declaring his independence from the Vatican."[28]

Obama is a Christian, and has been since leaving college. Yet, in the 2004 Senate race in Illinois, it was Obama's support of abortion in particular—and also gay marriage—that prompted his opponent, conservative Republican Alan Keyes, to assert that Jesus would not cast a vote for Obama. Obama later admitted that "Mr. Keyes's implicit accusation that I was not a true Christian nagged at me."[29]

Obama protested the charge, stating in his speech that there were certain issues that not only proved his Christianity but on which liberal Christian politicians generally could affirm their religiosity, if not turn the tables on conservative Christians—issues like supporting day care facilities and opposing the repeal of the estate tax. Such examples are commonly cited by liberal Christians; others include supporting a hike in the minimum wage, cleaning up the environment, and opposing tax cuts across the board—all part of the "social justice" mosaic of the religious left. There is, however, a major flaw in these pleas, frequently missed by liberal Christians:

Liberal and conservative Christians alike fully agree that Jesus wants them to help the poor. Yet they can easily, respectfully disagree over whether the estate tax or government-funded day care is what Jesus had in mind. Conservative Christians prefer to address poverty through individual outreach, nonprofit and faith-based organizations, and primarily the private sector; citing Scripture, they believe that Jesus pushed for private means of assistance. For example, the parable of the rich man getting into heaven calls not for a government program of forced wealth distribution but for the rich man to personally choose to share his wealth. Liberal Christians, on the other hand, favor public sector solutions, many of which their conservative counterparts find ineffective.

In short, this is a legitimate disagreement over means toward an agreed-upon end. Christian Democrats can scream in frustration over why conservative Christians will not vote for them as they uphold these social justice issues; as Michael Lerner has noted, there is even a tendency on the left to call such people on the right "stupid." Yet the intellectual flaw here is not in the mind of the conservative Christian. The mistake is to conclude that Jesus Christ would prefer an upper-income marginal tax rate of 36 percent instead of 31 percent. No Christian Democrat can claim to know that.

Similarly, all Christians agree that Jesus wants them to be good stewards of the earth, but no American politician can presume to know with absolute sureness whether Jesus would support the Kyoto Treaty or drilling for oil in the Persian Gulf but not in Alaska. All Christians agree it is wrong to discriminate against people on the basis of their race, but neither Democrat nor Republican can divine His precise will on quotas in admissions policy at the University of Michigan or on school vouchers in Milwaukee.

That said, there is a point where pro-choice Christian politicians like Obama and Mrs. Clinton reach a minefield: Conservative and liberal Christians alike agree that Jesus does not want them to kill the innocent. The disconnect between the two sides on abortion is over the view of the pro-life conservative Christian who cites the preeminence of the humanity of the unborn child, vs. the pro-choice liberal Christian who points to the preeminence of the mother's right to decide whether to terminate her pregnancy. That is the split, one which pro-life Christians feel is serious enough to make the pro-choice Christian at the least deeply confused and, according to some voices, like Alan Keyes's, even un-Christian; as a result, they will not cast a vote for a pro-choice Christian politician.

For many pro-life Christians, a legal abortion at twenty weeks for the purpose of birth control is not a matter of compromise, and is a far more significant barometer of one's Christian commitment than whether one advocates an increase in the minimum wage from $5.15 an hour to $5.45. They view the latter "social justice" concern as utterly insignificant by comparison.

This explains the statements of Alan Keyes toward Barack Obama. And Keyes, whom liberals feel is an extremist, represents a large segment of pro-life Christians. Such pro-lifers are not strong enough to lose Illinois or New York or California for Mrs. Clinton, but they can cost her Indiana, Missouri, the entire South, and sixty of Pennsylvania's sixty-seven counties.

Former Boston mayor Ray Flynn, committed to the Democratic Party to the grave and a pro-lifer whom President Bill Clinton appointed U.S. ambassador to the Vatican, talked to the *National*

Catholic Register about fellow Democrats like Obama (and Mrs. Clinton): "They talk like there's a big tent here in the Democratic Party. And then the next thing you know, when it's time for the political process to begin, they exclude pro-life Democrats like me." The Clintons in particular have been excluding pro-life Democrats since the 1992 convention, when they barred Pennsylvania Governor Bob Casey from speaking.[30]

Like Mrs. Clinton, Obama in his talk spoke of changing the way they talk about religion, but as Flynn noted, however, "It's not about rhetoric, it's about substance."[31] To think that rhetoric alone will win the day is to be guilty of the first of all sins: pride. Only pride could convince one that his or her opponents are so lacking in mental fiber as to be swayed by a shift in rhetoric but not substance. Mrs. Clinton should do what she did in her youthful days at Park Ridge when it came to civil rights: Take a ride to areas where she normally did not go and meet with people whom she was not like, dialogue with them, integrate with them, and not segregate herself in a way that allowed her to stereotype those on the other side in a simplistic way that demeaned them. After all, she did travel all the way to Calcutta to cooperate with a diminutive nun who told her that abortion was "evil" and that "life is the most beautiful gift of God."[32]

Matt Bai raised an intriguing possibility in the *New York Times Magazine*: that Hillary Rodham Clinton, if elected president, could catalyze a transformation within the Democratic Party, leading liberals into a kind of "new Democratic moralism," a "moral crusade." "Hillary [could] be the one," wrote Bai, "to remind baby-boom Democrats that most Americans believe in fixed points of right and wrong, both abroad and at home."[33] Could Hillary chart the way to a new Promised Land for Democrats, where the party sheds its well-earned image as the Secular Party?

That would indeed be quite a development. And she may make the effort. We can expect, for instance, a high-profile address by Mrs. Clinton on the subject of faith and politics well before November 2008, one aimed at attracting lots of publicity, which it would receive, and touted by an unthoughtful secular press that has suddenly found religion and no longer objects to fusing faith and politics, at least until the next George W. Bush comes along.

Again, however, such a transformation will require a shift in policy, not just rhetoric. Can the Democratic Party cast off the image of the Secular Party when it is increasingly perceived, as Ramesh Ponnuru has alluded, as the Party of Death, a party that supports not merely no limits on abortion but on embryonic research and euthanasia? That is the rub for Hillary.

Like the current president, Hillary Rodham Clinton is a big believer in intercessory prayer. "[N]ot only do I believe in it, I think there is increasing evidence of it," she says, pointing to the miraculous: "There is an interesting hospital study in which patients of comparable medical condition were prayed for, and prayers were, apparently, the only difference that could be discovered between how the patients were treated."[34]

As for herself personally, she says that she has "a lot of special prayers, and you know I rely on those in my daily life. . . . I carry a lot of them with me, but it's not something I really talk about. Except I would say this: There is just a real opportunity for people, through regular prayer and contemplation or just taking a few minutes out to think about themselves, to gain strength. And if it becomes a habit, it's always there for you. And I just hope more people, whatever their religious faith or spiritual beliefs might be, would try that. It can provide a great source of strength."[35]

Hillary looks for wisdom as a component of her Christian faith and personal growth. "When I went to Sunday school years ago our books often talked about how Jesus grew in wisdom and stature," she says. "I think about that often because it is unlikely I will grow any further in stature, but I certainly hope I will grow more in wisdom as the years go by."[36]

Time will only tell where such growth ultimately takes her, politically and spiritually. In the meantime, however, as she pursues the presidency, and seeks that middle-of-the-road voter as earnestly as her husband did in two successful presidential elections—and as she likely positions herself as the most religious Democratic front-runner since Jimmy Carter—it will be important that she be candid about where she stands as a Christian politician, and for the media to portray her accurately so the public can understand her correctly, even when it does not serve the interests of her supporters in the press.

Liberals and conservatives alike want clarity. Norman Lear, founder of the liberal group People for the American Way, is sick and tired of false advertising: "I love her," he said of Hillary. "But as terrific as I think she is, my concern is that we need someone who will tell the truth as they see it all of the time. She, like all of them, is not somebody who does that."[37]

Perhaps in this campaign Hillary's Christian compass that demands boldly proclaiming truth will override that Clintonian tendency to say certain things merely to get elected. That would not be out of the realm of Hillary Rodham Clinton's universe, especially for a woman never afraid to speak her mind, one who, though remaining a Methodist, has unfolded various wrinkles on that road from Hugh Rodham to Don Jones to Barry Goldwater to Bill Clinton, from Wellesley to Washington, from Michael Lerner to Jean Houston to Mother Teresa, from the stage of that Christmas pageant at Park Ridge Methodist to the pulpit of African-American congregations in Harlem.

And just as time will tell where that young woman from Park Ridge continues to move as a practicing Christian, time will also tell where voters—including those for whom faith is a central motivating factor—ultimately choose to move Hillary Rodham Clinton.

Acknowledgments

As always, there are many to thank for their part in this book. Chief among them are Cal Morgan and Judith Regan for requesting that I do more "God and . . ." books. Likewise, my agent, Leona Schecter, has encouraged the concept—and done much more. Also, at Harper-Collins, Matt Harper's thoughtfulness and superb editing had once again worked their magic.

In terms of research help, this book literally could not have been done without the assistance of Rachel Bovard, a brilliant young lady. She did an incredible amount of research in a short period of time during the 2005–2006 academic year, as she also simultaneously ran and acted in college plays and productions, headed the college honor society, aced her classes, and planned a wedding. (She is now Rachel Latta.) She also spared me the agony of two-finger typing lots of speech transcripts and statements from Bill and Hillary, and carefully ordered her research chronologically, a crucial step in organizing the book in a way that made the task of writing fluid and easy.

I took on this project at a time when I should not have, when I was overwhelmed with three book projects at various stages of progress—something I will never do again. If not for the fact that I was in my late thirties and in decent shape, I might not have survived the overload; it was overwhelming, requiring minute-by-minute management of my time for a two-year period. Rachel's hard work made this project possible.

Likewise extremely helpful on research was another exceptional young person, Stephen Albert, who picked up research tasks where Rachel left off after graduation. Playing likewise notable research roles were Marie Tyler, Leah Ayers, Andrew Larson, Betsy Christian, Sean Varner, Cory Shreckengost, Shawon Jackson-Ybarra, and Mitch and Paul Allan Kengor. All these individuals are current Grove City College students, recent graduates, or, in the case of Mitch and Paul, hopefully students-to-be someday. Among them, Marie Tyler's research on William F. Harrison was superb.

Grove City College is a special place. President Dick Jewell and Dean John Sparks, terrific leaders and fine men, are continuing to ensure that it remains special, as is Provost Bill Anderson, the "G.M.," who has assembled a team of remarkable faculty. All have been a wonderful source of support for the college and for me personally. Likewise a crucial source is Lee Wishing, who keeps the Center for Vision & Values moving as I juggle my directorship of the center along with my writing, teaching, fathering, husbanding, and all else. I also appreciate the diligence of Diane Grundy, Joyce Kebert, Conni Shaw, and the excellent Grove City College library staff.

I appreciate the love and support of family, from my parents to my wife, Susan, the latter of whom had serious reservations about my tackling this subject.

I'm also thankful for those two large, packed boxes of material on the Clintons that I had been clipping and saving and stockpiling since about 1992, for reasons I could never fathom. I twice almost threw those boxes in the trash. I'm glad I didn't.

Finally, my apologies to my friend and colleague Michael Coulter for keeping this book project secret from him, even on that spring day in 2006 when he entered my office and asked, as a gag for the annual "Faculty Follies" show at Grove City College, if I would stare into the video camera and, with a straight face, sarcastically declare that my next book would be *God and Hillary Clinton*. I declined the joke. Likewise in the dark was that wag who on the first day of classes in the fall of 2006 cut out a photo of Hillary and taped it to my office door. That prankster was on to something.

Notes

Preface

1. David Maraniss, *"First in His Class*: A Biography of Bill Clinton," Remarks from a forum at the Miller Center, University of Virginia, April 11, 1995.
2. In each faith-based biography I have done, a specific core issue, related to the subject's faith, emerges more prominently than any other, and which ignites the passions of the subject or the subject's critics. For Reagan, it was communism; for Bush, it was the Middle East, War on Terror, and the war in Iraq; for Hillary, it is abortion.
3. "Falwell Says Hillary Would Spark Base," Associated Press, September 24, 2006.

Chapter 1: Park Ridge Methodist

1. For more, see: Hillary Rodham Clinton, *Living History* (New York: Simon & Schuster, 2003).
2. For more, see: H. R. Clinton, *Living History*.
3. For more, see: Ibid.

4. Susan K. Flinn, ed., *Speaking of Hillary* (Ashland, Ore.: White Cloud Press, 2000), p. 14; excerpt from "Park Ridge: She Had to Put Up with Him," from Roger Morris, *Partners in Power: The Clintons and Their America* (Washington, D.C.: Regnery Publishing, 1996).

5. Gail Sheehy, *Hillary's Choice* (New York: Random House, 1999), pp. 22–23.

6. Ibid.

7. Ibid.

8. Ibid, pp. 24–25.

9. Barbara Olson, *Hell to Pay* (Washington, D.C.: Regnery Publishing, 1999), p. 29.

10. Hillary told this to Diane Huie Balay of the *Reporter,* a Methodist publication. The passage is cited in Peter Flaherty and Timothy Flaherty, *The First Lady* (Lafayette, La.: Vital Issues Press, 1995), pp. 20–21.

11. Flaherty and Flaherty, *First Lady*, p. 21.

12. Kenneth L. Woodward, "Soulful Matters," *Newsweek*, October 31, 1994.

13. Hillary Rodham Clinton, *It Takes a Village* (New York: Simon & Schuster, 1996), p. 171.

14. Ibid.

15. K. L. Woodward, "Soulful Matters"; and Hillary Rodham Clinton, Address to the 1996 United Methodist General Conference, April 24, 1996.

16. Hillary Rodham Clinton, Address to the 1996 United Methodist General Conference, April 24, 1996. Leon Osgood says of Dorothy's involvement at the church: "I remember Mrs. Rodham; she was not that active." See: Flaherty and Flaherty, *First Lady*, p. 21. To the contrary, most other accounts portray her as fairly active.

17. H. R. Clinton, Address to the 1996 United Methodist General Conference, April 24, 1996.

18. K. L. Woodward, "Soulful Matters."

19. Norman King, *Hillary: Her True Story* (New York: Birch Lane Press, 1993), p. 8.

20. H. R. Clinton, Address to the 1996 United Methodist General Conference, April 24, 1996.

21. Ibid.

22. Ibid.; Olson, *Hell to Pay*, pp. 29–31; and Hillary told this to Diane Huie Balay of the *Reporter,* a Methodist publication. The passage is cited in Flaherty and Flaherty, *First Lady*, pp. 20–21.

23. King, *Hillary*, p. 8.

24. Ibid.
25. Ibid.
26. H. R. Clinton, *It Takes a Village*, p. 171.

Chapter 2: The Don Jones Influence

1. One account had Jones at twenty-six years of age when he came to Park Ridge. Two others say he was thirty.
2. Barbara Olson, *Hell to Pay* (Washington, D.C.: Regnery Publishing, 1999), pp. 29–31.
3. Gail Sheehy, *Hillary's Choice* (New York: Random House, 1999), p. 35.
4. According to Gail Sheehy, he provided to the youth "a window into the more exotic worlds of abstract art, Beat poetry, existentialism, and the rumblings of radical political thought and counterculture politics that were eventually to explode under the smug slumber of even the good gray burghers of Park Ridge." Sheehy, *Hillary's Choice*, p. 33.
5. See: Roger Morris, *Partners in Power: The Clintons and Their Administration* (Washington, D.C.: Regnery Publishing, 1996), pp. 118–19.
6. "Park Ridge: She Had to Put Up with Him," from R. Morris, *Partners in Power*.
7. Norman King, *Hillary: Her True Story* (New York: Birch Lane Press, 1993), pp. 7–9.
8. Ibid.
9. Sheehy, *Hillary's Choice*, p. 33.
10. Christopher Andersen, *Bill and Hillary: The Marriage* (New York: William Morrow, 1999), pp. 95–97.
11. David Maraniss, *First in His Class* (New York: Simon & Schuster, 1995), pp. 251–52.
12. Peter Flaherty and Timothy Flaherty, *The First Lady* (Lafayette, La.: Vital Issues Press, 1995), pp. 20–25.
13. Sheehy, *Hillary's Choice*, p. 35.
14. Joyce Milton, *The First Partner* (New York: William Morrow, 1999), pp. 21–24.
15. Flaherty and Flaherty, *First Lady*, p. 23.
16. King, *Hillary*, pp. 7–9.
17. Interestingly, the source for many of these claims on the Web today is Gary Aldrich, the former Secret Service agent who worked for the Clintons in the 1990s and wrote a best-selling book on his experiences.

Aldrich's October 31, 2003, piece for *American Daily*, "The New Counterculture," has been widely reposted on the Web; that article, along with Aldrich's book and Web site, features the claim that Alinsky was a party member.

18. See: Saul D. Alinsky, *Rules for Radicals: A Practical Primer for Realistic Radicals* (New York: Random House, 1971), p. xiii.

19. Some biographers say that she wrote her thesis on Alinsky, whereas others say that she wrote on ideas that he shared, mentioning him only in passing. See, for example: Milton, *First Partner*, pp. 21–24; and R. Morris, *Partners in Power*, pp. 133–34.

20. Flaherty and Flaherty, *First Lady*, p. 21.

21. Milton, *First Partner*, pp. 21–24; and R. Morris, *Partners in Power*.

22. Joyce Milton reports that this situation made some parents in the congregation nervous, a problem of which Jones was oblivious at the time. "There was no flirting or anything like that," Jones says today. Milton, *First Partner*, pp. 21–24.

23. King, *Hillary*, pp. 7–9.

24. "Park Ridge: She Had to Put Up with Him," from R. Morris, *Partners in Power*.

25. Ibid.

26. Ibid.

27. Ibid.

28. Ibid.

29. Ibid.

30. King, *Hillary*, pp. 7–9.

31. Milton, *First Partner*, pp. 21–24.

32. Flaherty and Flaherty, *First Lady*, p. 24.

33. Ibid.

34. Ibid.

35. Ibid.

36. "Park Ridge: She Had to Put Up with Him," from R. Morris, *Partners in Power*.

37. As Peter Flaherty and Timothy Flaherty have noted, liberals have needed Don Jones to provide a flashpoint at which Hillary abandoned the alleged repression and closed-mindedness of middle-class suburbia, whereas conservatives have needed Jones as the political Svengali who transformed the Goldwater Girl into a left-wing radical. Most journalists, being liberal, have framed Don Jones's stay at Park Ridge in a positive light. Flaherty and Flaherty, *First Lady*, pp. 30–31.

Chapter 3: Hillary Hits Wellesley

1. Christopher Andersen, *American Evita: Hillary Clinton's Path to Power* (New York: HarperCollins, 2004), p. 23.

2. Judith Warner, *Hillary Clinton: The Inside Story* (New York: Signet Books, 1999), p. 31.

3. Charles Kenney, "Hillary: The Wellesley Years," *Boston Globe*, January 12, 1993.

4. Ibid.

5. Ibid.

6. Donnie Radcliffe, *Hillary Rodham Clinton: A First Lady for Our Time* (New York: Warner Books, 1993), p. 52.

7. Kenney, "Hillary: The Wellesley Years."

8. Ibid.

9. Joyce Milton, *The First Partner* (New York: William Morrow, 1999), p. 27.

10. Kenney, "Hillary: The Wellesley Years."

11. Roger Morris, *Partners in Power: The Clintons and Their Administration* (Washington, D.C.: Regnery Publishing, 1996), p. 127.

12. Kenney, "Hillary: The Wellesley Years."

13. David Maraniss, *First in His Class* (New York: Simon & Schuster, 1995), p. 255.

14. Kenney, "Hillary: The Wellesley Years."

15. Ibid.

16. Ibid.

17. Ibid.

18. Kenneth L. Woodward, "Soulful Matters." *Newsweek*, October 31, 1994

19. This has been reported by Ed Klein.

20. See: Wilson Yates, "*Motive* Magazine, the Student Movement, and the Arts," *Journal of Ecumenical Studies* 32, no. 4 (Fall 1995): 555–73.

21. Ibid., pp. 565–69.

22. K. L. Woodward, "Soulful Matters."

23. Carl Oglesby and Richard Shaull, *Containment and Change* (New York: Macmillan, 1967).

24. Cited by Laura Ingraham, *The Hillary Trap: Looking for Power in All the Wrong Places* (New York: Hyperion, 2000), p. 204.

25. Yates, "*Motive* Magazine," p. 567.

26. These opening words were recalled by editor Roy Eddey, who shared

them with Perry Brass. Brass was involved in the publication of this issue of *Motive*. Profile of Perry Brass, posted in the Religious Archives Network of the Lesbian, Gay, Bisexual, and Transgender Web site.

27. Ed Klein also lists some of these examples in his book, *The Truth About Hillary*. See also the review of Klein's book by Richard Poe, "The Woman Who Would Be President," FrontPageMagazine.com, July 5, 2005.

28. Yates, "*Motive* Magazine," p. 567.

29. Flaherty and Flaherty, *First Lady*, pp. 39–40.

30. Milton, *First Partner*, pp. 27–28.

31. Kenney, "Hillary: The Wellesley Years."

32. Ibid.

33. Gail Sheehy, *Hillary's Choice* (New York: Random House, 1999), p. 124.

34. Barbara Olson, *Hell to Pay* (Washington, D.C.: Regnery Publishing, 1999), p. 35.

35. Christopher Andersen, *Bill and Hillary: The Marriage* (New York: William Morrow, 1999), p. 98.

36. Kenney, "Hillary: The Wellesley Years"; also see Miriam Horn, *Rebels in White Gloves: Coming of Age with Hillary's Class* (New York: Times Books, 1998), pp. 46–47.

Chapter 4: God and Woman (and Bill) at Yale

1. Saul D. Alinsky, *Rules for Radicals: A Practical Primer for Realistic Radicals* (New York: Random House, 1971), pp. 18, 42.

2. Alinsky, *Rules for Radicals*. This feature quote from Alinsky appears after the dedication page and title page and immediately before the table of contents.

3. Roger Morris, *Partners in Power: The Clintons and Their Administration* (Washington, D.C.: Regnery Publishing, 1996), pp. 133–34.

4. Ibid., p. 48.

5. Ibid.

6. David Maraniss, *First in His Class* (New York: Simon & Schuster, 1995), p. 35.

7. Bill Clinton, *My Life* (New York: Random House, 2004), p. 30.

8. Maraniss, *First in His Class*, p. 35.

9. B. Clinton, *My Life*, p. 39.

10. Ibid.

11. R. Morris, *Partners in Power*, p. 49.

12. B. Clinton, *My Life*, p. 67.

13. By his own admission, Bill had not been a regular churchgoer since he left home for Georgetown in 1964, and had stopped singing in the church choir a few years before then. "After I went off to college," he confessed, "I became an erratic churchgoer." See: Ibid., p. 294; and Donald Baer, Matthew Cooper, and David Gergen, "Bill Clinton's Hidden Life," *U.S. News & World Report*, July 20, 1992.

14. Judith Warner, *Hillary Clinton: The Inside Story* (New York: Signet Books, 1999), p. 68.

15. Ibid., p. 31.

16. At the time, Medved was becoming interested in religion, and even attended some Quaker meetings. Yet he did not personally observe "any interest in religion on her [Hillary's] part, or any conversation, or any expression of religiosity at all." Medved says that "for months" he had initially thought that Hillary might be Jewish. Interviews with Michael Medved, June 8 and 9, 2006.

17. Paul Lewis, "Robert Treuhaft, Lawyer Who Inspired Funeral Expose, Dies at 89," *New York Times*, December 2, 2001.

18. See: Ibid.; Rick DelVecchio, "Robert Treuhaft, Crusading Bay Area Lawyer, Champion of Leftist Causes for Decades," *San Francisco Chronicle*, November 12, 2001.

19. DelVecchio, "Robert Treuhaft."

20. Andersen reports it as a menorah. It would have been a mezuzah instead of a menorah; a menorah is a candelabrum, whereas a mezuzah is an enclosed scroll to mark the doorpost of a home. Christopher Andersen, *American Evita: Hillary Clinton's Path to Power* (New York: HarperCollins, 2004), p. 50.

21. Ibid.

22. There was speculation to this effect by several Arkansas pro-life leaders interviewed for this book.

23. See: Hillary quoted in Claire G. Osborne, ed., *The Unique Voice of Hillary Rodham Clinton: A Portrait in Her Own Words* (New York: Avon Books, 1997), p. 18; and Norman King, *Hillary: Her True Story* (New York: Birch Lane Press, 1993), p. 60.

24. I am not familiar with a single source that documents the churches (or homes) in which all first couples were married. Research on more recent presidents, however, confirms that the Trumans, Kennedys, Fords, Carters, and all the Bushes were married in churches, with information on the others requiring more research. The Eisenhowers, however, appear

to have been married in Mamie's family home, which was probably the result of frictions over the denomination of Ike's mother. In short, nearly all the more recent first couples were married in churches.

25. B. Clinton, *My Life*, p. 370; and King, *Hillary*, p. 60.
26. Andersen, *American Evita*, p. 60.
27. Peter Flaherty and Timothy Flaherty, *The First Lady* (Lafayette, La.: Vital Issues Press, 1995), pp. 75–76.
28. Hillary Rodham Clinton, *Living History* (New York: Simon & Schuster, 1996), pp. 72–73.

Chapter 5: The First Lady of Arkansas

1. Peter Flaherty and Timothy Flaherty, *The First Lady* (Lafayette, La.: Vital Issues Press, 1995), p. 139.
2. Ibid., p. 141.
3. Joyce Milton, *The First Partner* (New York: William Morrow, 1999), p. 140.
4. David Maraniss, *First in His Class* (New York: Simon & Schuster, 1995), pp. 432–33.
5. Maraniss agrees, seeing a similarity in the Clintons' religious evolutions, reflective of a broader generational trend: Both saw churchgoing as an essential part of their early adolescent years, less apparent in their twenties, and then, as Maraniss describes it, "more vital again" as they moved into their thirties and into parenthood and middle age. Ibid.
6. Ibid.
7. Ibid.
8. Information taken from the Web site of the First United Methodist Church of Little Rock.
9. Bishop Richard B. Wilke said this in his introductory remarks on behalf of Hillary Rodham Clinton in her April 24, 1996, speech to the United Methodist General Conference. Also see: Maraniss, *First in His Class*, pp. 432–33.
10. During a July 2006 phone call placed to the church in the course of researching this book, a church secretary said that services had been televised since at least the late 1960s. According to the timeline posted on the church Web site, services began to be televised in 1973.
11. The perception is so commonplace that there is no need for a citation. In addition to Nigel Hamilton, *Bill Clinton: An American Journey: Great*

Expectations (New York: Random House, 2003), among the more careful biographers who at least note the perception is Milton, *First Partner*, p. 140.

12. This is according to the Web site of the Southern Baptist Convention.

13. Bill Clinton, *My Life* (New York: Random House, 2004), p. 294.

14. Donald Baer, Matthew Cooper, and David Gergen, "Bill Clinton's Hidden Life," *U.S. News & World Report*, July 20, 1992.

15. B. Clinton, *My Life*, p. 294.

16. Hamilton, *Bill Clinton*, pp. 379–80.

17. Baer, Cooper, and Gergen, "Bill Clinton's Hidden Life."

18. B. Clinton, *My Life*, p. 294.

19. Baer, Cooper, and Gergen, "Bill Clinton's Hidden Life."

20. Maraniss, *First in His Class*, p. 434.

21. Ibid.

22. Ibid., pp. 434–35.

23. Interview with William F. Harrison, January 10, 2007.

24. Maraniss, *First in His Class*, pp. 434–35.

25. Ibid.

26. Ibid.

27. Depending, of course, on the translation, some Bibles use the words "spirit" or even "wind" rather than "breath of life," and some use the word "formed" instead of "fashioned."

28. Some other sources, including Norman King, *Hillary: Her True Story* (New York: Birch Lane Press, 1993), date the visit in 1980. Clinton himself says December 1981.

29. King, *Hillary*, pp. 84–85.

30. B. Clinton, *My Life*, p. 294.

31. Ibid.

32. Ibid.

33. King, *Hillary*, pp. 84–85.

34. Ibid.

35. Milton, *First Partner*, p. 139.

36. Yet their sudden appearances in church, concedes Milton, "may not have been as calculated as it appeared." Milton, too, noticed other aspects of the Clintons' religious life that suggest sincerity rather than politics. Milton, *First Partner*, pp. 139–40.

37. Maraniss does not provide the date or place or occasion for the speech, though Hillary gave the speech when her husband was governor. Maraniss, *First in His Class*, pp. 432–33.

38. Ibid.

39. Interview with Pastor Ed Matthews, October 25, 2005.

40. Maraniss, *First in His Class*, pp. 432–33.

41. Interview with Pastor Ed Matthews, October 25, 2005.

42. Though Lewis recalls Hillary teaching, he does not recall her attending, though he says that his wife's memory is "infinitely better" than his, and that she recalls Hillary also attending as well as teaching.

43. Interview with Willard Lewis, November 3, 2005.

44. Interview with Willard Lewis, November 3, 2005.

45. Others from Hillary's Sunday school class refused to respond to requests for interviews, and advised Lewis thereafter to do the same. When asked on November 7, 2005, if there was anyone else in the class that he suggested be contacted for this book, Lewis said that he forwarded the request to Craig and Nancy Wood, both longtime members of FUMC and the Bowen-Cabe Sunday school class, of which Craig is now president. Nancy, a former teacher, had served as a trustee of Hendrix College and as a member of the state board of education. Lewis that said both were "very substantial and civic-minded individuals in whose word you can have absolute trust." Neither of these individuals responded. In response to a follow-up inquiry made directly to Lewis on February 14, 2006, Lewis wrote: "I had mentioned to Nancy Wood that I had had some inquiries about Hillary. Her response, and a rather emphatic one too, was that she scrupulously avoided responding to any inquiries about the Clintons, out of concern that the information thus accumulated might be put to some negative use. Sorry I couldn't be of more help!"

46. Wilke, introductory remarks, April 24, 1996.

47. Attendees in this class did not want to be interviewed for this book, with the exception of Willard Lewis.

48. Interview with Willard Lewis, November 7, 2005.

Chapter 6: Hillary's Causes and Bill's Demons

1. Nigel Hamilton, *Bill Clinton* (New York: Random House, 2003), p. 467.

2. Ibid.

3. Joyce Milton, *The First Partner* (New York: William Morrow, 1999), p. 158.

4. Ibid.

5. See: Herbert J. Gans, "The Uses of Poverty: The Poor Pay All," *Social Policy* (July/August 1971): 20–24.

6. Milton, *First Partner*, pp. 158–60.

7. Ibid.

8. Ibid., p. 159.

9. Ibid., p. 158–60.

10. Judith Warner, *Hillary Clinton: The Inside Story* (New York: Signet Books, 1999), p. 270.

11. Gail Sheehy, *Hillary's Choice* (New York: Random House, 1999), p. 173.

12. Ibid.

13. The source for this anecdote is Christopher Andersen, *American Evita: Hillary Clinton's Path to Power* (New York: HarperCollins, 2004), p. 60. Norman King says that the reception took place the next day at Ann Henry's "place." King, *Hillary: Her True Story* (New York: Birch Lane Press, 1995), p. 60. In *My Life*, Bill Clinton seems to imply that the reception took place the evening following the ceremony, where he says "a couple hundred of our friends" gathered at Ann Henry's house before they shifted to "Billie Schneider's place in the Downtown Motor Inn" where they all "danced the night away"; Clinton, *My Life* (New York: Random House, 2004), pp. 233–35.

14. Interview with Pastor Ed Matthews, October 25, 2005.

15. Interviews with Pastor Ed Matthews, October 25 and 31, 2005.

16. Interviews with Pastor Ed Matthews, October 25 and 31, 2005.

17. Sheehy, *Hillary's Choice*, p. 190.

18. Ibid.

19. King, *Hillary*, p. 124.

20. For example, Christopher Andersen writes only: "The Methodist marriage counselor apparently worked wonders; Hillary emerged after several emotional sessions to proclaim that she and Bill had recommitted themselves to the marriage." It is not totally clear if the "Methodist minister" is Ed Matthews, and Andersen does not list a source. Andersen, *American Evita*, p. 101.

21. Martin Walker, *The President We Deserve: Bill Clinton: His Rise, Falls, and Comebacks* (New York: Crown Publishers, 1996), pp. 113–14.

22. Ibid.

23. Sheehy, *Hillary's Choice*, p. 189.

24. Ibid., p. 190.

25. Warner, *Hillary Clinton*, pp. 275–76.

26. Sheehy, *Hillary's Choice*, p. 189.

27. Walker, *President We Deserve*, p. 114.

28. Milton, *First Partner*, p. 194.

29. Warner, *Hillary Clinton*, pp. 274–75.

30. Interview with Pastor Ed Matthews, October 25, 2005.

31. Interview with Pastor Ed Matthews, October 25, 2005.

32. Interview with Pastor Ed Matthews, October 25, 2005.

33. Hillary Rodham Clinton, *It Takes A Village* (New York: Simon & Schuster, 1996), p. 151.

34. On this, also see: Roxanne Roberts, "16 Candles for Chelsea," *Washington Post*, February 27, 1996; and Roxanne Roberts, "Life with Father," *Washington Post*, January 27, 1998.

35. See: Andersen, *American Evita*, pp. 92–93.

Chapter 7: Taking Power

1. Norman King, *Hillary: Her True Story* (New York: Birch Lane Press, 1993), p. 7.

2. Interviews with Pastor Ed Matthews, October 25 and 31, 2005.

3. Donald Baer, Matthew Cooper, and David Gergen, "Bill Clinton's Hidden Life." *U.S. News & World Report*, July 20, 1992.

4. Peter Flaherty and Timothy Flaherty, *The First Lady* (Lafayette, La.: Vital Issues Press, 1995), p. 155.

5. Frank Bruni, "Senior Bush's Loss Set Course for Son's Candidacy," *New York Times*, December 26, 1999.

6. Flaherty and Flaherty, *First Lady*, p. 152.

7. See: Meredith Oakley, *On the Make: The Rise of Bill Clinton* (Washington, D.C.: Regnery Publishing, 1994), p. 97. Author cites an interview with Jones published by *People* magazine, January 25, 1993.

8. Bill Clinton, *My Life* (New York: Random House, 2004), p. 474.

9. On this, see: George Weigel, *Witness to Hope: The Biography of Pope John Paul II* (New York: HarperCollins, 2001), p. 715.

10. Kenneth L. Woodward, "Soulful Matters," *Newsweek*, October 31, 1994.

11. B. Clinton, *My Life*, p. 563.

12. K. L. Woodward, "Soulful Matters."

13. Interviews with Philip Wogaman, October 13 and 24, 2005. On where the first family sat, the source is K. L. Woodward, "Soulful Matters."

14. Interviews with Philip Wogaman, October 13 and 24, 2005.

15. Interviews with Philip Wogaman, October 13 and 24, 2005.

16. Hillary Rodham Clinton, Address to the 1996 United Methodist General Conference, April 24, 1996.

17. K. L. Woodward, "Soulful Matters."

18. Senator Al Gore, *Earth in the Balance: Ecology and the Human Spirit* (New York: Houghton-Mifflin, 1992), pp. 163, 238–65, 269, 272–74, 282–83, 293–94. Also see: Paul Kengor, *Wreath Layer or Policy Player? The Vice President's Role in Foreign Policy* (Lanham, Md.: Lexington Books, 2000), pp. 241–45.

19. Susan K. Flinn, ed. *Speaking of Hillary* (Ashland, Ore.: White Cloud Press, 2000) pp. 79–88; speech given as part of Liz Carpenter Lecture Series, 1993.

20. Michael Kelly, "Saint Hillary," *New York Times Magazine*, May 23, 1993.

21. Hillary said this. See: Ibid.

22. Gail Sheehy, *Hillary's Choice* (New York: Random House: 1999), p. 234.

23. "The Rodham Family Biography," CNN.com.

24. Kelly, "Saint Hillary."

25. Ibid.

26. Ibid.

27. Gail Sheehy, pp. 234–35.

28. Ibid.

29. Martha Sherrill, *Washington Post*, "Hillary Clinton's Inner Politics," May 6, 1993.

30. Ibid.

31. Joyce Milton, *The First Partner* (New York: William Morrow, 1999), p. 284.

32. Kelly, "Saint Hillary."

33. Henry Allen, "A New Phrase at the White House," *Washington Post*, June 9, 1993.

34. Ibid.

35. Ibid.

36. Michael Lerner, "Hillary's Politics, My Meaning," *Washington Post*, June 13, 1993.

37. See: Barbara Olson, *Hell to Pay* (Washington, D.C.: Regnery Publishing, 1999), pp. 312–13; and Milton, *First Partner*, p. 283–84.

38. Allen, "New Phrase at the White House."

39. Ibid.

40. Ibid.

41. Ibid.

42. Ibid.

Chapter 8: The Clintons, the Pope, and Mother Teresa

1. George Weigel, *Witness to Hope: The Biography of Pope John Paul II* (New York: HarperCollins, 2001), p. 681.
2. Transcript of August 12, 1993, remarks by Pope John Paul II. Also see: Weigel, *Witness to Hope*, p. 681.
3. *Evangelium Vitae* was issued in March 1995.
4. For the record, some, like George W. Bush, would later see Communism's ugly successor as radical Islam, marked by a "turning point," as Bush called it, on September 11, 2001. In fact, both the pope and Bush could be correct, in that radical Islam and uncontrolled abortion and liberalization of embryonic research and euthanasia could all have the effect of killing innocents—all create a Culture of Death, whether in a skyscraper in Manhattan, a café in Tel Aviv, or the white rooms of a laboratory, hospital, or abortion clinic.

   Later, Haynes Johnson, a sympathetic voice, would castigate Clinton for not seizing the moment, for, in essence, wasting his chance in the 1990s to grab this transition by the horns and define a new era. Of course, Johnson, being a liberal, did not have in mind what the pope was talking about. And he could not have had in mind what Bush was talking about; neither did Bush nor anyone else, as no one expected September 11, 2001, an event that not only transformed the course of world history but the entirety of what otherwise would have been a fairly mundane Bush presidency focused on a few domestic issues stemming from his "compassionate conservatism."
5. Weigel, *Witness to Hope*, p. 684.
6. See: Robert P. Casey, *Fighting for Life* (Nashville, Tenn.: Thomas Nelson, 1996), p. 175.
7. Raymond Hernandez and Patrick D. Healy, "The Evolution of Hillary Clinton," *New York Times,* July 13, 2005.
8. Weddington said that he was "not proposing that you send federal agents armed with Depo-Provera dart guns to the ghetto. You should use persuasion rather than coercion."
9. This information, including a full copy of the Weddington letter to Clinton, was exposed in a special report by the legal watchdog Judicial Watch. The report is titled "The Clinton RU-486 Files," and is available

on the Web site of Judicial Watch.

10. Clinton notes that he attended each one of them. I was not able to determine if Hillary attended each, and Bill did not say so. On Bill, see: Bill
Clinton, *My Life* (New York: Random House, 2004), p. 558.

11. Kathryn Spink, *Mother Teresa, A Complete Authorized Biography* (San Francisco: HarperSanFrancisco, 1997), p. 272.

12. Hillary Rodham Clinton, *Living History* (New York: Simon & Schuster,
2003), pp. 417–18.

13. A text of the speech is available on the Web site of Eternal Word Television Network, EWTN.

14. Henry Allen, "A New Phrase at the White House," *Washington Post*,
June 9, 1993.

15. Peggy Noonan, "Still, Small, Voice," *Crisis*, February 1998, pp. 12–
17.

16. Ibid.

17. Spink, *Mother Teresa*, p. 272.

18. Hillary spoke later that afternoon at the National Prayer Luncheon.

19. H. R. Clinton, *Living History*, pp. 417–18.

20. Spink, *Mother Teresa*, p. 272; and Mary McGrory, "Allies for the Children," *Washington Post,* June 20, 1995.

21. McGrory, "Allies for the Children"; and H. R. Clinton, *Living History*,
p. 418.

22. H. R. Clinton, *Living History*, pp. 417–18.

23. McGrory, "Allies for the Children"; and H. R. Clinton, *Living History*,
p. 418.

24. H. R. Clinton, *Living History*, pp. 417–18.

25. Ibid., p. 418; and Spink, *Mother Teresa*, pp. 283–84.

26. Gail Sheehy, *Hillary's Choice* (New York: Random House, 1999),
pp. 234–35.

Chapter 9: The Debacle of November 1994

1. Michael Lerner, "Hillary's Politics, My Meaning," *Washington Post*, June
13, 1993.

2. The only major news story I could find on this issue was a February
23, 2001, report by Andrea Peyser in the conservative *New York Post*,
reprinted on the conservative Web site FreeRepublic.com.

3. See: Alan Cowell, "Vatican Says Gore Is Misrepresenting Population

Talks," *New York Times*, September 1, 1994, p. A1.

4. Georrge Weigel, *Witness to Hope: The Biography of Pope John Paul II* (New York: HarperCollins, 2001) pp. 724–25.

5. Kenneth L. Woodward, "Soulful Matters," Newsweek, October 31, 1994.

6. Ibid.

7. Ibid.

8. Ibid.

9. Ibid.

10. Ibid.

11. Her husband states: "Most ardent pro-lifers are all for prosecuting doctors, but grow less certain when their argument that an abortion is a crime is carried to its logical conclusion: prosecuting the mother for murder." See: Bill Clinton, *My Life* (New York: Random House, 2004), p. 229. Former California Supreme Court justice William P. Clark, who in 1981 declined Ronald Reagan's offer of a seat on the U.S. Supreme Court, notes that no woman in the history of American jurisprudence has been prosecuted for having an abortion—doctors have been, as have midwives, but not the mother. "And pro-lifers don't want that now," says Clark, today a major player in California's Propositions 73/85 ballot initiative for parental notification for abortion laws. "Compassion has always been shown to the mother." Interview with William P. Clark, July 3, 2006.

12. Hillary Rodham Clinton, *Living History* (New York: Simon & Schuster, 2003), p. 418.

13. K. L. Woodward, "Soulful Matters."

14. Gail Sheehy, *Hillary's Choice* (New York: Random House, 1999), p. 244.

Chapter 10: New Agers and Eleanor's Ghost

1. Marian Wright Edelman, "Protect Children from Unjust Policies," *Washington Post*, November 3, 1995.

2. Peter Edelman, "The Worst Thing Bill Clinton Has Done," *Atlantic Monthly*, March 1997, pp. 43–58.

3. Peter Flaherty and Timothy Flaherty, *The First Lady* (Lafayette, La.: Vital Issues Press, 1995), p. 212.

4. Text of First Lady Hillary Rodham Clinton's remarks at the Annual

National Prayer Luncheon, February 2, 1995.

5. Bob Woodward, *The Choice* (New York: Simon & Schuster, 1996), pp. 55–57.

6. Ibid., p. 56.

7. Ibid.

8. Hillary Rodham Clinton, "Remarks for the United Nations Fourth World Conference on Women," Beijing, China, September 5, 1995.

9. This seems to have been misreported in a ZENIT story four years later, titled "Hillary Clinton Calls Abortion a 'Human Right,'" published by ZENIT on February 9, 1999.

10. B. Woodward, *Choice*, p. 271–72.

11. Ibid., p. 129.

12. Ibid.

13. Ibid., pp. 130–32.

14. Ibid., pp. 132–33.

15. Ibid.

16. Ibid., p. 412.

17. Barbara Olson, *Hell to Pay* (Washington, D.C.: Regnery Publishing, 1999), pp. 313–14.

18. Jon Klimo, *Channeling: Investigations on Receiving Information from Paranormal Sources* (Los Angeles: Jeremy P. Tarcher, 1987), pp. 188–200.

19. Claire Osborne, *The Unique Voice of Hillary Rodham Clinton: A Portrait in Her Own Words* (New York: Avon Books, 1997), p. 121.

20. Gail Sheehy, *Hillary's Choice* (New York: Random House, 1999), p. 262. For the quote, Sheehy cites the *Los Angeles Times* of June 25, 1996.

21. I was not able to ascertain if she had sought out this speaking opportunity or had been invited without prompting.

22. H. R. Clinton, Address to the 1996 United Methodist General Conference, April 24, 1996.

23. Osborne, *Unique Voice of Hillary Rodham Clinton*, pp. 89–90.

Chapter 11: Surviving the Second Term

1. Polling data posted by *Meet the Press with Tim Russert*, NBC, June 28, 1998.

2. For example, see: Lois Romano and Peter Baker, "Another Clinton Accuser Goes Public," *Washington Post*, February 20, 1999, p. A1. Broaddrick herself would confront Hillary with a letter to the sitting first lady

demanding an explanation, a letter excerpted in the press. See, among others: Steve Miller, "Broaddrick Confronts First Lady in Angry Letter," *Washington Times*, October 17, 2000.

3. Barbara Olson saw something deeper in Hillary's warnings of a dark conspiracy against her husband, something rooted in the moral absolutism of her religious worldview. "It became the root of her worldview, one in which it is never enough to attack an opponent's actions," wrote Olson. "One must also expose his motives, and use that perspective to destroy both the action and its proponents. For the natural companion of a doctrine of perfectibility is a conviction in the existence of evil—and immorality—of one's enemies. Hillary's America is a starkly Manichean universe, one in which she perceives the enemies of progress as numerous, powerful, and clever." Olson, *Hell to Pay* (Washington, D.C.: Regnery Publishing, 1999), p. 30.

4. Jonathan Alter, "Why Hillary Still Holds On," *Newsweek*, August 31, 1998.

5. Maraniss interviewed by Tom Brokaw, NBC coverage of the speech, August 17, 1998.

6. Beschloss and Stephanopoulos provided commentary on ABC immediately after the speech, August 17, 1998.

7. Chris Matthews speaking on MSNBC's *Hardball*, February 12, 1999.

8. See: Robert D. McFadden, John Kifner, and N. R. Kleinfeld, "10 Days in the White House: Public Acts and Secret Trysts," *New York Times*, September 14, 1998; and Nancy Gibbs, "The Starr Report: We, the Jury," *Time*, September 21, 1998.

9. The show aired on ABC's *20/20* on March 3, 1999. Also see: Editorial, "It's Fun!" *Washington Times*, March 5, 1999.

10. Alter, "Why Hillary Still Holds On."

11. Ibid.

12. Ibid.

13. Ibid.

14. Ibid.

15. Christopher Andersen, *Bill and Hillary: The Marriage* (New York:William Morrow, 1999), p. 32.

16. The fact that Jackson invited himself has been widely reported in the various biographies of Hillary.

17. Jesse Jackson, "Keeping Faith in a Storm," *Newsweek*, August 31, 1998.

18. Gail Sheehy, *Hillary's Choice* (New York: Random House, 1999), p. 311; and Andersen, *Bill and Hillary*, pp. 24–25.

19. Andersen, *Bill and Hillary*, pp. 24–25.

20. Ibid.
21. Jackson, "Keeping Faith in a Storm."
22. Ibid.
23. See, among others, Christopher Andersen, *American Evita: Hillary Clinton's Path to Power* (New York: HarperCollins, 2004), pp. 167–68.
24. Jackson, "Keeping Faith in a Storm."
25. Andersen, *Bill and Hillary*, pp. 24–25.
26. Jackson, "Keeping Faith in a Storm."
27. Ibid.
28. See: Sheehy, *Hillary's Choice*, p. 311; and Andersen, *Bill and Hillary*, pp. 24–25.
29. Andersen, *Bill and Hillary*, pp. 24–25.
30. Of the three, only MacDonald did not respond to requests to be interviewed. The request was made through his receptionist on October 18 and November 2, 2005.
31. On this, see: Bob Woodward, *Shadow: Five Presidents and the Legacy of Watergate* (New York: Touchstone, 1999), p. 324.
32. Andersen, *Bill and Hillary*, p. 31.
33. Bill Clinton, *My Life* (New York: Random House, 2004), p. 810.
34. Interviews with Philip Wogaman, October 13 and 24, 2005.
35. Interview with Tony Campolo, March 22, 2006.
36. Interview with Tony Campolo, March 22, 2006.
37. Interviews with Philip Wogaman, October 13 and 24, 2005.
38. Andersen, *Bill and Hillary*, p. 32.
39. Interviews with Philip Wogaman, October 13 and 24, 2005.
40. Interview with Tony Campolo, March 22, 2006.
41. Andersen, *Bill and Hillary*, p. 31.
42. Interviews with Philip Wogaman, October 13 and 24, 2005.
43. See: "Chronology of Clinton's Apologies," *New York Times*, September 11, 1998.
44. Ibid.
45. B. Clinton, *My Life*, pp. 810–11.
46. As Hillary attended Immanuel Baptist, Hillary sat in her regular pew at First United Methodist when the head usher whispered to her asking if she would be willing to receive the offering, and she happily agreed. The offering comes after the sermon, at the end of the service. Says Matthews: "She knew her cue as to when she was supposed to do that, and moved back to the end of the church. Security people [the Secret Service] had no idea what she was doing, and actually, neither did I! But she loved being able to do that for her church. To be a servant." Interview with Pastor Ed

Matthews, October 25, 2005.

47. Sheehy, *Hillary's Choice*, pp. 322–23.

48. Interviews with Pastor Ed Matthews, October 25 and 31, 2005.

49. Vincent Morris, "First Churchgoers Get That Old-Time Religion," *Washington Post,* September 21, 1998.

50. B. Clinton, *My Life*, pp. 811, 846.

51. On that, Judith Warner states that "Hillary has repeatedly stressed that her religious beliefs, as much as anything else, have borne her through her marriage." But was there something more than her religion? How about her politics? Warner asks, "Is religion really the key behind Hillary Rodham Clinton's decision to stay with Bill Clinton? In an abstract sense, yes." Warner explains that Hillary's faith demands that she "sublimate" her personal feelings to a greater, abstract ideal. "Life can have some transcendent meaning," Hillary once said in a graduation speech at Hendrix College. "Work toward the achievement of a universal human dignity, not just your own personal security." Warner speculated that Hillary appears to have made a conscious effort to focus on the transcendent aspects of her marriage. She says that Hillary has always had faith in Bill Clinton's larger political mandate, that she "passionately" believed in the good he could do for the country. To accomplish that mandate, reported Warner, Hillary concluded that Bill needed her beside him, both as a "wifely presence" and as a political adviser. Warner is quick to add that politics was not Hillary's only consideration: "She also, by all accounts, has always loved him." See: Judith Warner, *Hillary Clinton: The Inside Story* (New York: Signet Books, 1993), p. 276.

52. Quoted in Zev Chafets, "When It Comes to Faith, This Senator's Full of It," *Daily News,* June 9, 2003.

53. Laura Ingraham, *The Hillary Trap: Looking for Power in All the Wrong Places* (New York: Hyperion, 2000), p. 197.

54. Ibid.

55. Howard Kurtz, "A Reporter with Lust in Her Hearts," *Washington Post,* July 6, 1998.

56. Cited in L. Brent Bozell III, "Bigwigs of 'Primary Colors' Offer Insights on Libido and Leadership," *Pittsburgh Tribune-Review*, March 26, 1998.

57. Ibid.

58. Dana Gresh speaking on James Dobson's *Focus on the Family* daily radio broadcast, December 18, 2006.

59. Cited in Charles E. Dunn, *The Scarlet Thread of Scandal* (Lanham, Md.: Rowman-Littlefield, 1999), p. 1.

60. See: B. Woodward, *Shadow*, pp. 451, 474–75; and Woodward inter-

viewed on *Tim Russert*, CNBC, June 20, 1999.

61. Quoted in B. Woodward, *Shadow*, p. 475.

62. Andrew Johnson, the other president to be impeached, had come into office as Lincoln's successor after Lincoln was assassinated. Johnson had been Lincoln's vice president.

63. For a complete list, see: "Cincinnati Paper Blasts Clinton," MSNBC-News.com, September 17, 1998.

64. On this, see: George Weigel, *Witness to Hope: The Biography of Pope John Paul II* (New York: HarperCollins, 2001), pp. 740, 776–77.

65. Julia Lieblich, "Pope Arrives in U.S.," Associated Press, July 27, 1999; and "Pope Arrives in St. Louis and Warns of 'Time of Testing,'" CNN.com, July 27, 1999.

66. Ibid.

67. "President Clinton's Meeting with His Holiness Pope John Paul II," Statement by the Press Secretary, the White House, January 26, 1999.

Chapter 12: Transition

1. Her comments, said the *New York Times*, were the first time she had publicly discussed her views on this issue. Adam Nagourney, "Hillary Clinton Faults Policy of 'Don't Ask,'" *New York Times*, December 9, 1999.

2. Ibid.

3. Ibid.

4. Gail Sheehy, *Hillary's Choice* (New York: Random House, 1999), pp. 151–52.

5. Joel Siegel, "Hil Nixes Same-Sex Marriage," *Daily News,* January 11, 2000.

6. Ibid.

7. Beth Harpaz, "Hillary Clinton Booed at Parade," Associated Press, March 18, 2000.

8. Ibid.

9. "Hillary Marches in St Patrick's Day Parade," Catholic League News Release, March 17, 2006.

10. She would instead be marching in Syracuse's local St. Patrick's Day parade. Fellow Democrat and former New York City mayor Ed Koch called her excuse "ridiculous." Bob Kappstatter of the *Daily News* said the decision would not win her an award for a profile in courage. See: "Hillary Marches in St Patrick's Day Parade," Catholic League News

Release, March 17, 2006; and Bob Kappstatter, "Hil: I'm Not Marching in Parade," *Daily News*, March 9, 2001.

11. Adam Nagourney, "Hillary Clinton Vows to Fight to Preserve Abortion Rights," *New York Times*, January 22, 2000.

12. Ibid.

13. Ibid.

14. Marc Humbert, "Hillary Clinton Scoring on Abortion Rights Front," Associated Press, August 9, 2000.

15. See: Suzanne Fields, "First Anti-Semite?" *Washington Times,* July 24, 2000.

16. Christopher Andersen, *American Evita: Hillary Clinton's Path to Power* (New York: HarperCollins, 2004), pp. 187–88.

17. See: Beth Harpaz, *The Girls in the Van: Covering Hillary* (New York: St. Martin's Press, 2001), pp. 117–35.

18. Ibid.

19. Ibid.

20. Ibid., pp. 186–87.

21. Ibid.

22. Ibid.

23. Ibid., p. 188.

24. Ibid., pp. 173–74.

25. Ibid., p. 195.

26. Ibid., p. 198.

27. Ibid., pp. 198–99.

28. Ibid., pp. 173–74.

29. Ibid., pp. 173–74.

30. Adam Nagourney, "Mrs. Clinton Preaches to the Party Faithful," *New York Times*, November 6, 2000, pp. A1 and B5. Also see: Maggie Haberman, "Mass Appeal as 'Sister Hillary Tours Churches,'" *New York Post*, November 6, 2000, p. 7.

31. Ibid.

32. Also see: Steve Miller, "Hillary Courts Blacks at Church Services," *Washington Times*, November 6, 2000.

33. Polling data posted by *Meet the Press with Tim Russert*, NBC, June 28, 1998.

34. On this, see the extended discussion in Paul Kengor, *God and George W. Bush: A Spiritual Life* (New York: Regan Books, 2004), pp. 80–83.

35. President Bill Clinton, "Remarks to African American Religious Leaders," New York City, October 31, 2000.

36. President Bill Clinton, "Remarks to the Congregation of Alfred Street

Baptist Church," Alexandria, Va., October 29, 2000.

37. President Bill Clinton, "Remarks to the Congregation of Shiloh Baptist Church," Washington, D.C., October 29, 2000.

38. For more examples, see the lengthy treatment of this subject in: Kengor, *God and George W. Bush*.

39. President Bill Clinton, "Remarks at the New Hope Baptist Church," Newark, New Jersey, October 20, 1996.

40. Four came in 2000, six in 1996, and two in 1994.

41. One was a memorial service in Landover, Maryland, honoring Martin Luther King Jr. The other came in Atlanta in June 2002. Another came on January 15, 2004, in Atlanta. This does not include memorial services like the September 11 service at the National Cathedral, to which he was invited, expected, and happy to come, and was joined there by the Clintons. The Atlanta appearance did feature a plug for his faith-based programs. President George W. Bush, "Remarks at St. Paul A.M.E. Church," Atlanta, Georgia, June 21, 2002.

42. See my discussion of this in Kengor, *God and George W. Bush*.

43. Katharine Q. Seelye and Kevin Sack, "Gore Rallies Base," *New York Times*, November 6, 2000.

44. Ann McFeatters, "Head to Head, Neck and Neck," *Pittsburgh Post-Gazette*, November 5, 2000.

45. James O'Toole, "Candidates Blitz Pittsburgh," *Pittsburgh Post-Gazette*, November 5, 2000; Ralph Z. Hallow, "Gore Used Race as His Ace Card in Election," *Washington Times*, November 9, 2000; and Marisol Bello, "Gore Appeals to Pittsburgh's Black Voters," *Pittsburgh Tribune-Review*, November 5, 2000.

Chapter 13: Senator Clinton

1. Data provided by the Federal Election Commission.

2. Christopher Andersen, *American Evita: Hillary Clinton's Path to Power* (New York: HarperCollins, 2004), pp. 231–32.

3. See: Bill Sammon, *Fighting Back* (Washington, D.C.: Regnery Publishing, 2002), pp. 159–60; and Bob Woodward, *Bush at War* (New York: Simon & Schuster, 2003), pp. 66–67.

4. For numerous such examples, see chapter 12 of Paul Kengor, *God and George W. Bush: A Spiritual Life* (New York: Regan Books, 2004),

pp. 219–49.

5. Raymond Hernandez and Patrick D. Healy, "The Evolution of Hillary Clinton," *New York Times*, July 13, 2005.

6. "Hillary's Senate Record," *Washington Times,* November 21, 2004.

7. Ibid.

8. Hernandez and Healy, "Evolution of Hillary Clinton."

9. "Hillary's Senate Record," *Washington Times,* November 21, 2004.

10. Dean told reporters: "I am still learning a lot about faith and the South and how important it is." The *Boston Globe* reported: "[Dean] said he expects to increasingly include references to Jesus and God in his speeches as he stumps the South." Sarah Schweitzer, "Seeking a New Emphasis, Dean Touts His Christianity," *Boston Globe*, December 25, 2003. Also see: Franklin Foer, "Beyond Belief: Howard Dean's Religion Problem," *The New Republic*, December 29, 2003; William Safire, "Job and Dean," *New York Times*, January 5, 2004; Ted Olsen, "'Allegory' Job 'Favorite Book in the New Testament,' Says Howard Dean," ChristianityToday.com, January 5, 2004; David Teather, "Democrat Hopeful Walks with God in 'Bible Belt,'" *The Guardian*, January 5, 2004; Jim VandeHei, "Dean Now Willing to Discuss His Faith," *Washington Post*, January 4, 2004; and Jodi Wilgoren, "Dean Narrowing His Separation of Church and Stump," *New York Times*, January 4, 2004.

11. Michael Goodwin, "Howard and Hillary Sing the Same Tune," *New York Daily News*, June 12, 2005.

12. See: Hernandez and Healy, "Evolution of Hillary Clinton"; and Marc Humbert, "Hillary Clinton Scoring on Abortion Rights Front," Associated Press, August 9, 2000.

13. Clinton-Lazio Debate, held in Manhattan, October 8, 2000.

14. Ralph Z. Hallow, "Carter Condemns Abortion Culture," *Washington Times*, November 4, 2005.

15. Clinton-Lazio Debate, held in Manhattan, October 8, 2000.

16. Clinton issued vetoes on April 10, 1996, and October 10, 1997. "Former President Clinton Insults Pope on Way to Funeral," LifeSiteNews.com, April 7, 2005.

17. See: Daniel H. Johnson Jr., M.D., President, American Medical Association, Letter to the Editor, *New York Times*, May 26, 1997, p. A22.

18. It was one of these exemptions that effectively made abortion legal in California under Governor Ronald Reagan, who had no idea that what he signed into law would be so abused.

19. This was the 1996 Christmas Eve service.

20. Interview with Rob Schenck, November 6, 2006.

21. Schenck says that several Secret Service agents searched him. When he asked them why they had apprehended him, one of the senior agents made a motion toward President Clinton and remarked, "He told us to." Schenck then pulled out his cell phone to call his attorney; as he did, one of the younger agents smacked the phone from his hand, knocking it to the floor and accidentally activating the "send" button, which called Schenck's office. As it dialed Schenck's office, his answering machine was able to record the interrogation that followed. According to Schenck, the White House denied being involved in the incident. To dispute this denial, Schenck played the audio recording for a reporter who had heard about the incident and approached him. The White House then dropped the issue.

22. The bill was next sent to the House of Representatives. Final Senate approval of the vote came on October 21, 2003, which gave approval to the final version of the bill (called a "conference report"). This vote was upped to sixty-four to thirty-four, and then went to President Bush for his signature.

23. See: Editorial, "Don't Legalize Gay Marriage," *Washington Times,* June 25, 2003.

24. Deborah Orin, "Dems Do a Tiptoe Around Gay Marriage," *New York Post,* July 3, 2003.

25. Andrew Sullivan, "Clinton's Cowardice," *New York Sun,* June 20, 2003.

26. Ibid.

27. The "right to privacy" was manufactured by a majority of U.S. Supreme Court justices in the 1965 case of *Griswold v. Connecticut*, in which the Court spoke of "penumbras" and emanations from other rights, such as the right against unreasonable search and seizure. These penumbras or "glows" from other rights were regarded by the Court majority as creating this new right to privacy. Justices Black and Stewart issued stinging dissents on this issue.

28. Hillary Rodham Clinton, *It Takes a Village* (New York: Simon & Schuster, 1996), p. 178.

29. Lisa Bennett, "Over One Million March for Women's Lives," *National NOW Times,* Spring 2004.

30. This material and that which follows is taken from "Abortion Rights Protest Packs National Mall," Associated Press, April 25, 2004; Kathryn Jean Lopez, "We're . . . Feminists!" *National Review Online,* April 26, 2004; Kathryn Jean Lopez, "Wax Bush," *National Review Online,* April

26, 2004; George Neumayr, "Among the Pagan Ladies," *The American Spectator*, April 26, 2004; and Nicole Colson, "A March for Women's Lives or Democrats' Votes?" *Socialist Worker Online*, April 30, 2004.

31. Hillary Frey, "Marching for Women's Lives," *The Nation*, April 26, 2004.

32. Colson, "March for Women's Lives or Democrats' Votes?"

33. "Hillary Clinton Speaks Out at March," CapitalNews9.com, April 25, 2004.

34. Colson, "March for Women's Lives or Democrats' Votes?"

35. Frey, "Marching for Women's Lives."

36. "Hillary Clinton Speaks Out at March," CapitalNews9.com, April 25, 2004; and Neumayr, "Among the Pagan Ladies."

37. Kerry said this on the Senate floor on August 2, 1994.

38. CNN posted these data the day after the voting finished.

39. I know this personally because I spoke at length to the *Daily News* reporter, who called me for comment on the Rove strategy to mobilize churches. While commenting, I made repeated references to what Clinton had done the day before at Riverside. The reporter was unaware of what happened at Riverside and seemed totally uninterested.

40. David R. Guarino, "Hill at Tufts: Use the Bible to Guide Poverty Policy," *Boston Herald*, November 11, 2004.

41. "Hillary's Poll Numbers Startling," Fox News, December 19, 2004.

Chapter 14: Moving to the Middle

1. "Senator Clinton Blasts Bush Administration over Abortion," *Catholic Exchange*, January 18, 2005.

2. "Sen. Clinton on Government Role in Caring for Sick," Canisius College Press Release, January 18, 2005.

3. "Catholic Groups Boycott Hillary Speech," *Newsmax*, January 29, 2005.

4. "Canisius College Invites Hillary Clinton, Abortion Views Upset Catholics," *Life News*, February 1, 2005.

5. "Senator Clinton Blasts Bush Administration over Abortion," *Catholic Exchange*, January 18, 2005.

6. "Catholic Groups Boycott Hillary Speech," *Newsmax*, January 29, 2005.

7. "Bishop Withdraws from Hillary Speech," *Newsmax*, January 30, 2005.

8. See: Michael Jonas, "Sen. Clinton Urges Use of Faith-based Initiatives," *Boston Globe*, January 20, 2005; and Joe Mariani, "Hillary Clinton: One

Step Back, Shift to the Right," *American Daily,* January 28, 2005.

9. Editorial, "Hillary Tunes Up," *New York Sun*, January 25, 2005.

10. Editorial, "Senator Clinton's Values Lesson," *New York Times*, January 30, 2005.

11. Joseph Curl, "Hillary in the Middle on Values Issues," *Washington Times*, January 26, 2005.

12. Quoted in George Neumayr, "Safe, Legal, and Hillary," *The American Spectator*, January 27, 2005.

13. Randy Hall, "Catholic Group Opposes College's Plan to Honor Senator Clinton," *The Nation*, April 21, 2005.

14. Ibid.

15. Ibid.

16. Also see: Joyce Purnick, "Passing Buck on Schiavo Cheats Public," *New York Times*, August 28, 2006.

17. The Ivins piece was posted on the Common Dreams Web site on January 20, 2006.

18. "Clintons: Pope a 'Beacon' for Democracy," *Newsmax*, April 3, 2005.

19. To its credit, the conservative Web source NewsMax.com reported Clinton's remarks in full on April 8, 2005.

20. "Former President Clinton Insults Pope on Way to Funeral," LifeSite-News.com, April 7, 2005.

21. Michael J. Gaynor, "Pope John Paul II Is Praying for Bill Clinton," MichiganNews.com, April 10, 2005.

22. Brian Williams, "Bill Clinton Remembers Pope John Paul II," MSNBC.com, April 7, 2005; and Bill Sammon, "Bush Keeps Low Profile at John Paul II's Funeral," *Washington Times*, April 8, 2005.

23. "Christian Defense Coalition: Senator Clinton Snubs Dialogue with Pro-Life Groups," Press Release, Christian Defense Coalition, April 25, 2005.

24. Bill Clinton, *My Life* (New York: Random House, 2004), p. 558.

25. Hillary Rodham Clinton, *It Takes a Village* (New York: Simon & Schuster, 1996), pp. 174–78.

26. Kristen Lombardi, "Time for a Prayer Circle," *Village Voice*, April 12, 2005.

27. The press release from the college came on April 20. Mrs. Clinton had voted on the Mexico City Policy on April 5. Hall, "Catholic Group Opposes College's Plan to Honor Senator Clinton."

28. Tim Drake, "Marymount Manhattan: Catholic or Not?" *National Catholic Register*, May 22–28, 2005, p. 11; and Hall, "Catholic Group Opposes

College's Plan to Honor Senator Clinton."

29. Hall, "Catholic Group Opposes College's Plan to Honor Senator Clinton."

30. Ibid.

31. Ibid.

32. Ibid.

33. Ibid.

34. To cite one, see: "Catholic Groups Boycott Hillary Speech," *Newsmax*, January 29, 2005. The terminology was also used by *National Review*, the *National Catholic Register*, and others.

35. Raymond Hernandez and Patrick D. Healy, "The Evolution of Hillary Clinton," *New York Times*, July 13, 2005.

36. This occurred on August 30, 2005. See: Michael McAuliff, "Hil Shreds FDA Boss on Abort Pill," *New York Daily News*, August 31, 2005.

37. Clinton-Lazio Debate, held in Manhattan, October 8, 2000.

38. "Clinton's 'Plantation' Remark Draws Fire," CNN.com, January 19, 2006.

39. Ibid.

40. Ibid.

41. Ibid.

42. Ibid.

43. Michael McAuliff and Helen Kennedy, "Hil Has a Holy Cow over Immigrant Bill," New York *Daily News*, March 23, 2006.

44. For the record, Bush has been at odds with his own party through much of the immigration debate, and has generally not toed a hard line.

45. McAuliff and Kennedy, "Hil Has a Holy Cow over Immigrant Bill."

46. "Hillary Clinton: Right Wingers to Blame for Abortion," NewsMax. com, May 19, 2006.

47. For the Focus on the Family news article on Dobson's statements, see: Gary Schneeberger, "Dobson Challenges Media to Fairly Report Stem-Cell Issue," CitizenLink.org, June 28, 2004.

Chapter 15: Hillary and the Faith Factor

1. Bill scored at 7 percent and 43 percent, respectively. The other figures measured were Pat Robertson, Pat Buchanan, George W. Bush, and Al Gore, who were ranked in order from most to least religious. See accompanying story: Elizabeth Crowley, "Bible Gives Shape to Conservative

Values of Fundamentalist Voters," *Wall Street Journal*, March 9, 2000.

2. Comments were made on January 20, 2001, with Sidey interviewed by Larry King and Caddell by Chris Matthews.

3. Among others, see Dowd's "She Wants to Babaloo Too," *New York Times*, September 8, 1999; and "Dragon Lady Politics," *New York Times*, October 13, 1999.

4. Susan Page, "Poll Majority Say They'd Be Likely to Vote for Clinton," *USA Today*, May 27, 2005.

5. Klein is the author of *The Truth About Hillary*. He said this in "The Truth About Hillary," Q&A by Kathryn Jean Lopez, National Review Online, June 20, 2005; and on the radio talk show, "Quinn in the Morning," WPGB, 104.7 FM, Pittsburgh.

6. Christopher Flickinger, "Author Ed Klein Claims Clintons 'Sold Their Souls to the Devil in Order to Achieve Power,'" Human Events Online, July 8, 2005.

7. "Hillary Marches in St Patrick's Day Parade," Catholic League News Release, March 17, 2006.

8. Ibid.

9. "Frontline: American Porn," *Frontline*, February 7, 2002 (original airdate), Program #2012, Transcript, p. 11.

10. Ibid.

11. Michael Kelly, "The Flynt Standard," *Washington Post*, January 13, 1999.

12. "Frontline: American Porn," *Frontline*, February 7, 2002 (original airdate), Program #2012, Transcript, p. 14.

13. See, for instance, the 2004 *Book of Discipline* for the United Methodist Church. The 2004 edition is the most recent edition at the time of the writing of this book. The next edition will come out in 2008.

14. See the Web site for the Religious Coalition for Reproductive Choice.

15. The *Book of Discipline* of the United Methodist Church states unequivocally: "[W]e support the legal option of abortion under proper medical procedures." Among the contradictions in the *Book of Discipline* is this statement on abortion: "We cannot affirm abortion as an acceptable means of birth control." In fact, the vast majority of abortions are for the purpose of birth control.

16. Jennifer Senior, "The Once and Future President Clinton," *New York Magazine*, November 15, 2005.

17. Bill Clinton, *My Life* (New York: Random House, 2004), p. 229.

18. Ibid.

19. Hillary Rodham Clinton quotes on "Religion" (undated), compiled in Claire G. Osborne, *The Unique Voice of Hillary Rodham Clinton: A Portrait*

in Her Own Words (New York: Avon Books. 1997), pp. 88–90.

20. Raymond Hernandez and Patrick D. Healy, "Evolution of Hillary Clinton," *New York Times*, July 13, 2005.

21. "'What Does Hillary Think?' New Special Report Raises Disturbing Questions about Hillary Clinton and Abortion," Judicial Watch, November 27, 2006. The May 1993 Galston memo and others are posted at the Web site of Judicial Watch.

22. This was reported by Fred Barnes, "A Pro-Life White House," *The Weekly Standard*, January 1/8, 2001, p. 13.

23. Interview with William F. Harrison, January 10, 2007; and Stephanie Simon, "Offering Abortion, Rebirth," *Los Angeles Times*, November 29, 2005.

24. Steven Ertelt, "Democratic Prez Candidates Target Pro-Life Voters with Religious Outreach," LifeNews.com, December 13, 2006.

25. The information on Carter in this section is taken from Ralph Z. Hallow, "Carter Condemns Abortion Culture," *Washington Times*, November 4, 2005.

26. "Zogby/Associated Television News Poll Reveals: Abortion Tough Issue for Hillary Clinton & '06 Congressional Democrats," Associated Television News / PR Newswire, March 22, 2006.

27. Tom McFeely, "Democrats Woo Religious," *National Catholic Register*, August 6–12, 2006.

28. E. J. Dionne Jr., "Obama's Eloquent Faith," *Washington Post*, June 30, 2006.

29. McFeely, "Democrats Woo Religious."

30. Ibid.

31. Ibid.

32. Mother Teresa wrote this in her official letters to Mrs. Clinton and the other delegates at the United Nations Fourth World Conference on Women in Beijing in September 1995, and to the Cairo conference a year earlier.

33. Matt Bai, "Mrs. Triangulation," *New York Times Magazine*, October 2, 2005.

34. Hillary Rodham Clinton quotes on "Religion" (undated), compiled in Claire G. Osborne, *The Unique Voice of Hillary Rodham Clinton: A Portrait in Her Own Words* (New York: Avon Books, 1997), pp. 88–90.

35. Ibid.

36. Ibid., pp. 88–90.

37. Hernandez and Healy, "Evolution of Hillary Clinton."

Index

CPSIA information can be obtained at www.ICGtesting.com
Printed in the USA
LVOW04s1500180915

454634LV00026B/226/P